Toward fields covered with recent stars
Across harvests of stars and reflections
Of the fire of four horseshoes splashing the veils
He dives to the very depths of the milky darkness

Unfurling the ribbon of the abolished cycles,
The shortest ones bending under the weight of the sunsets
Since, as suns whose light had grown pale, they had
Come too close to the redness of the Lyre and of Hercules

But, at this hour, the moon in her bridal dress
Drags at her white heels the nebula and white
White as the morning upon the petrified sea
The ram of dawn prepares to set out

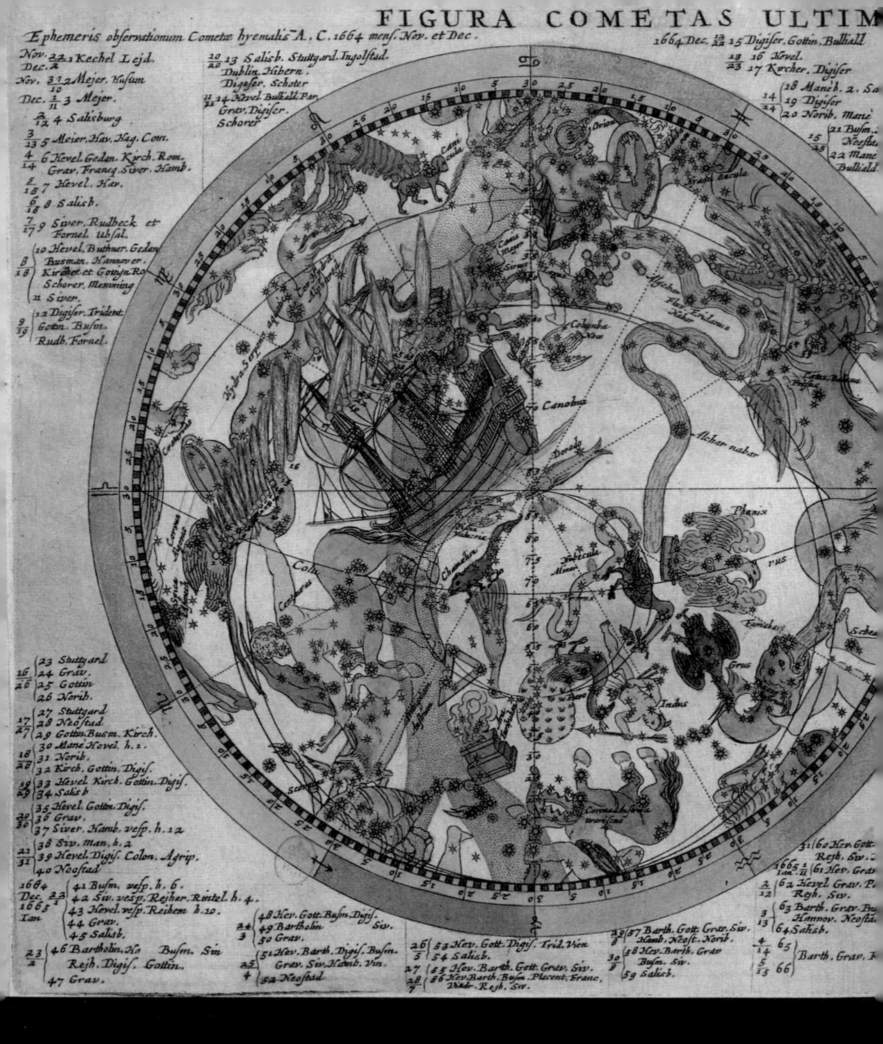

The comet has placed its sparks on his forehead
Beautiful black woman, oh moon, where are you going so slowly
To find your spouse with his plum-colored eyes
Whose bed Venus warmed with a gallant body?

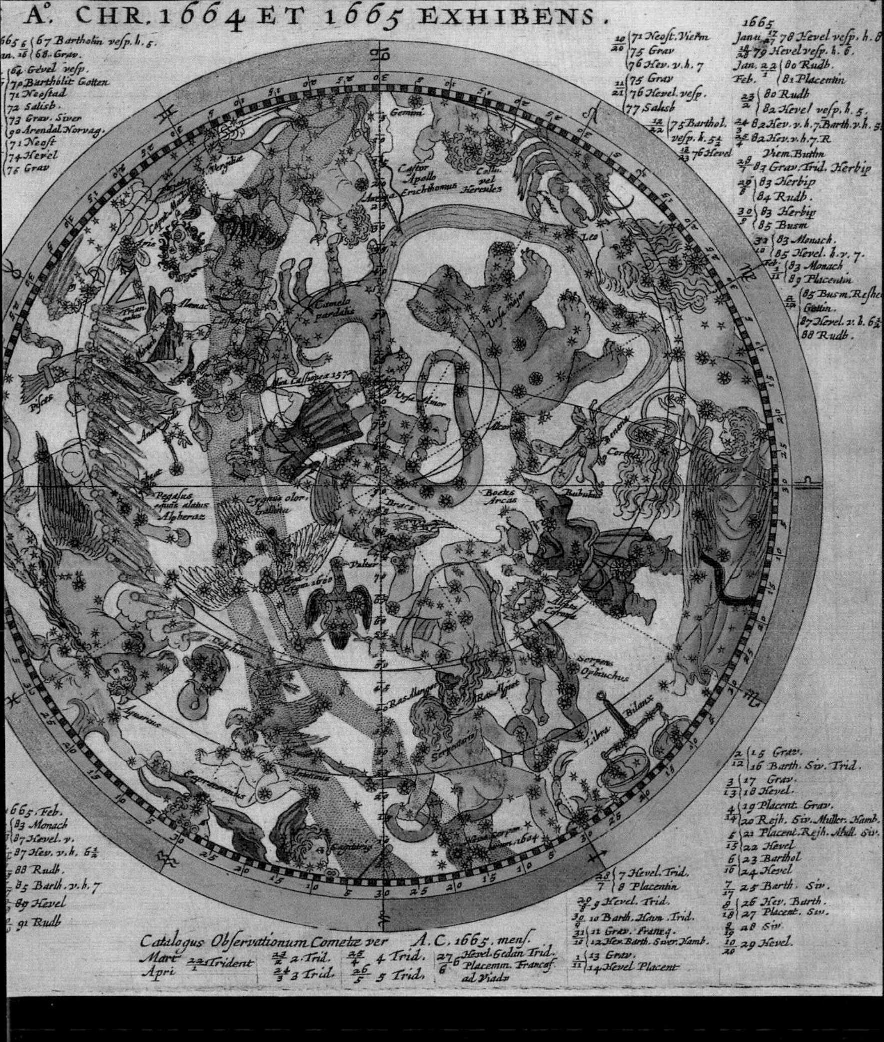

Champagne, flow through the constellations
If wines are like liquid stars
Let us find, in you, Burgundy, the creation
Of the fabulous monsters of the ether and of the void

Pressing the grapes we shall conjure up
Mercury and Jupiter and Cancer and Ursa Minor
Despite the torches reflected in the wine
And the sun bathed in the coolness of its springs

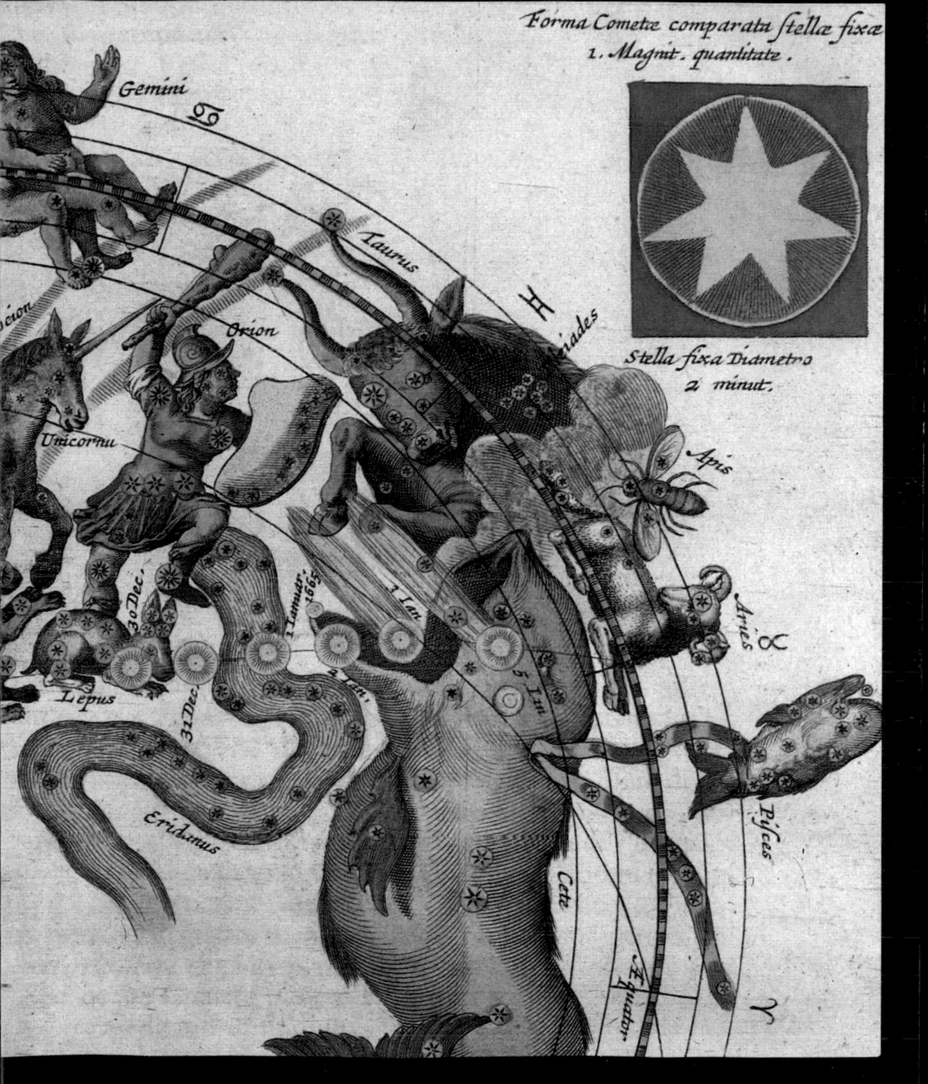

You, fine after-midnight, escorted by legends
Drag one more couple into the waltzes of desire
So that the wearied drinker may again ask you
To fill his glass with the blood of memories.

Robert Desnos, *The Fine After-Midnight*

CONTENTS

THE SKY
MYSTERY, MAGIC, AND MYTH

Jean-Pierre Verdet

DISCOVERIES
HARRY N. ABRAMS, INC., PUBLISHERS
NEW YORK

The universe is nothing more than a great void, dotted with stars that often appear in immense concentrations, called galaxies. Among the stars float enormous clouds of gas and dust. More than five billion years ago, one of those great cosmic clouds began to collapse in on itself, giving birth to the star we call the sun and to its chain of planets—Mercury, Venus, Earth, Mars, Jupiter, Saturn, Uranus, Neptune, and Pluto—and their satellites, or moons, which together make up our solar system.

CHAPTER I
THE SKY OF HUMANKIND

The colored engraving at left was published in 1647 in Johannes Hevelius' *Selenografia sive Lunae Descriptio.* Right: According to 16th-century historian and theologian Conrad Lycosthenes, in 1157 two moons appeared in the sky, with a sun in the middle marked by a cross, and he represented that apocryphal event in an engraving.

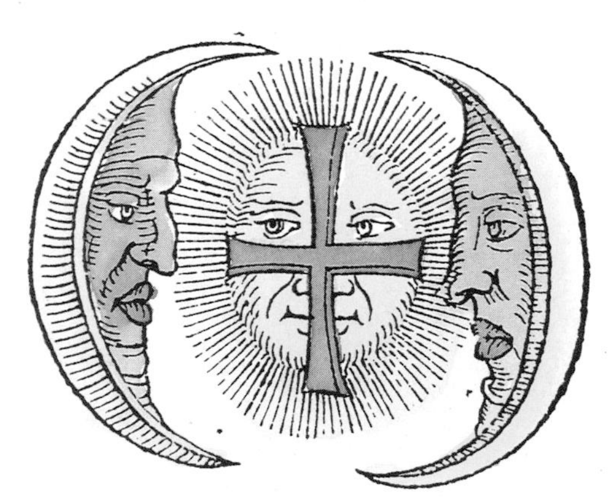

Originating as molten magma, Earth eventually cooled down and solidified. For a long time matter alone reigned over a mineral and aquatic world. Then, around two billion years ago, in a molecular form that could barely be distinguished from inert matter, life appeared, gradually adapting, developing, and growing more complex until it finally established itself on the continents that had emerged from the ocean.

Scientists have divided the known geological development of Earth into five major eras; the current Cenozoic era began about sixty million years ago. The eras are further divided into periods and subperiods. By the end of the Tertiary period, about one and a half million years ago, the terrestrial landscape was similar to the one we are familiar with, with the same geography and the same flora and fauna. There was

These enormous stones at Stonehenge, in England, were arranged in a circle almost four thousand years ago. Beside the main group stands a large solitary stone. Each year on around 22 June, in the morning, a person standing in the middle of the circle can see the sun rise directly behind the large single stone, a phenomenon that has led some people to believe the monument was dedicated to the sun.

only one significant absence, that of human beings (although paleontologists believe that "protohumans" were already present in a race of quadrupeds). While they have yet to trace fully the progression of development of the human race, scientists do know that it was early in the Ice Age, which followed the Tertiary period, that the first humans to stand upright made their appearance.

For more than a million years, one hears nothing of humankind. Human beings left very little evidence of their technical achievements, and even though those artifacts that have been found—sharpened pebbles, tools made of stone fragments, and scrapers—presuppose a degree of intelligence, there is no material evidence testifying to their mental activity.

This carved stone tablet, called a stela, from Rocher-des-Doms near Avignon, France, is more than five thousand years old. Inscribed on its face is a sun with eight rays coming from it.

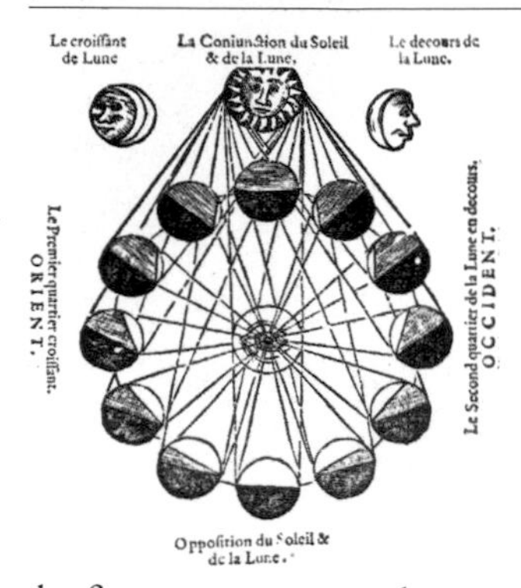

It was only in the last part of the Stone Age, approximately fifty thousand years ago, that evidence of human expression appeared: bones decorated with red pigment, small balls of limestone adorned with bits of stone and fragments of bone, the first engraved stones, the first attempts at sculpture, the first tombs. On the earliest engraved stones one can make out groups of stars, or constellations, indicating that even prehistoric humans fixed their gaze upon the sky.

The combination of the moon's orbit around Earth and Earth's orbit around the sun determines the phases of the moon, which are identified by the portions of the moon visible from Earth during the period of a month. One half of the moon is always illuminated by the sun, a phenomenon seen only when the moon is full. The size of the illuminated portion of the disk that we see changes in the course of the moon's rotation around Earth. This engraving of the phases of the moon (left) was published in 1524 by the 16th-century German astronomer Peter Bennewitz, known as Apianus, in *Cosmographia sive Descriptio Universi Orbis.*

From the Dawn of Time Humans Have Observed Unchanging Regularities in the Cosmos: Days, Seasons, and Phases of the Moon

From the moment they first connected the alternation of night and day with the motion of the heavens, human beings have been fascinated by astronomy, or the science of celestial bodies—their origins, their composition, their movements. Long before they knew how to write, humans knew the phases of the moon, which formed the basis of the first calendars. They noticed the periodic return of the seasons. They watched the daily movement of the stars, which they saw as circular and uniform, and observed every night that the stars reappeared in the sky in what seemed to be permanent arrangements; they found it useful to group the stars into constellations and sometimes to label them according to the images they saw in them. Scholars believe that even from prehistoric times the study of the sky has stimulated a dual development in human thought: the search for natural and unchanging laws by which events, such as those that occur

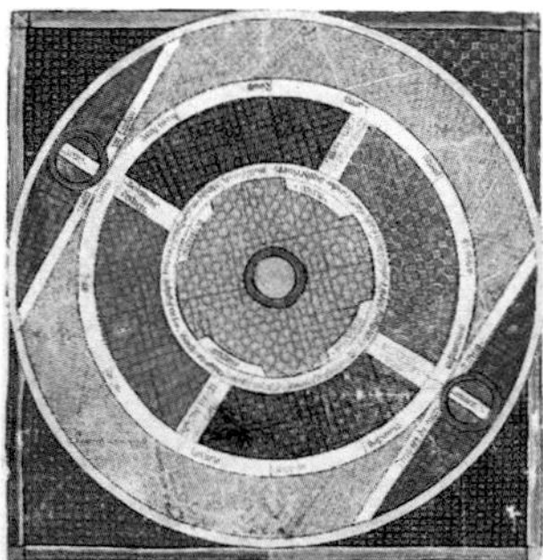

in or derive from the sky, could be understood and thereby planned for, and the temptation to place in the sky, which then appeared inaccessible, omnipotent beings to explain many of those mysterious and sometimes frightening events.

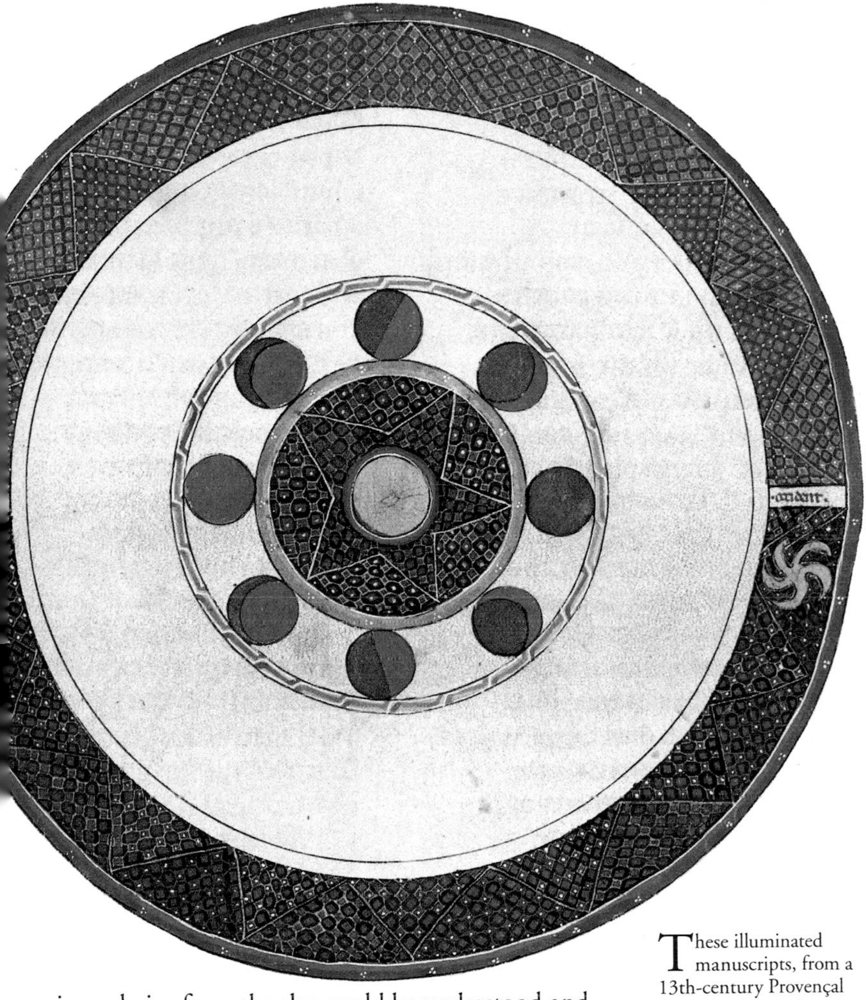

Links between certain celestial and earthly phenomena were recognized from an early period. For

These illuminated manuscripts, from a 13th-century Provençal breviary of love, chart the progress of the sun through the zodiac (opposite) and the phases of the moon (above); the portion of the moon visible from Earth is in red.

example, the phases of the moon could be correlated with the schedule of the tides, and the succession of the seasons with the sun's progress through the stars. Evidence of specific, labeled constellations of stars—the Bull, the Scorpion, and the Lion—can be traced to the Euphrates Valley around 4000 BC. It was much later, probably sometime in the 5th century BC, that the ecliptic—the apparent path of the sun through the heavens during the course of a calendar year—was identified in Babylon, and perhaps in Greece also. The Babylonians took all the stars that fell across the ecliptic and divided them into the twelve constellations known today as the signs of the zodiac. They studied the positions of these stars in relation to other celestial bodies in order to determine the wishes of their gods and assure the well-being of their kings, and they

Among the celestial wonders recorded in this 18th-century engraving are comets (in the form of swords), flaming beams of light, shooting stars, and fire-causing lightning. The scene in the upper right reflects the contemporary belief that great social upheavals were preceded by aerial combat between horsemen.

developed stories, or myths, that connected occurrences in the sky with the activities of the gods. The study of the influence of the stars and planets on human affairs became known as astrology.

Order and Disorder, Those Fundamental Opposites, Appeared in Both Scientific and Popular Studies of the Sky

Early peoples must have been fascinated also by the movement of the planets against the great network of stars. Along with the major periodic astronomical phenomena there occurred exceptional, often spectacular events, such as solar or lunar eclipses or the appearance of comets or shooting stars. The sky overhead could also grow angry, producing thunder and lightning, flooding rains, and violent windstorms such as hurricanes or tornadoes. Thus, along with the order that controlled so many constant and apparently eternal events there existed an unpredictable and often terrifying disorder. The duality is reflected in many creationary myths that show the universe emerging out of chaos, returning to disorder, and bursting forth again, regenerated and reordered.

Before the appearance of astronomical telescopes at the beginning of the 17th century, astronomers could rely only on very basic instruments. A common example is shown in this 14th-century bronze relief by Andrea Pisano; the instrument, in the shape of a quarter circle, with its simple line of sight and lead wire, enabled observers to determine the position of the stars above the horizon.

Human Imagination and Intelligence Conceived of a Sky That Was Truly Multifaceted

From the time that humans first looked upward until December of 1609, when the Italian astronomer and philosopher Galileo Galilei pointed his specially made telescope toward the stars, all people were equal when observing the sky. All they had to rely on were their eyes and their intelligence. From simply gazing at the sky, some developed a science that they considered exact, even before they really had the means to form such a science. Others developed empirical rules concerning agriculture, navigation, or the forecasting of the weather; still others worked out myths that often developed into legends and folklore practices.

It was predominantly in relation to themselves that people thought when contemplating the sky, and the

Michael Wolgemut's wood engraving of 1491 illustrates an episode from the biblical book of Joshua relating a battle between the Hebrews and the Canaanites, during which Joshua commanded: "'Sun, stand thou still at Gibeon, and thou moon in the valley of Aijalon!' And the sun stood still and the moon stayed, until the nation took vengeance on their enemies. Is this not written in the book of [the Just One]? The sun stayed in the midst of heaven and did not hasten to go down for about a whole day. There has been no day like it before or since, when the Lord hearkened to the voice of a man."

Joshua 10:12–4

cosmogonical myths speak above all of humankind. The Roman scholar Pliny the Elder looked at the sky and observed such mythological characters as the Bear, the Bull, Perseus, the Northern Crown, and Berenice's Hair. Pliny was amazed that some people imagined the outermost sphere of the heavens to be smooth, whereas he saw a sky rich in imagery, encrusted with the forms of every animal and every object on Earth.

Early Theories About the Sky Have Been Preserved in Myths

This book does not presume to give a complete account of the scientific study of the sky or even to describe the early stages of that science and its subsequent evolution. Its aim is to present a selection of popularly held ideas about the sky and its manifestations and to explore the symbols and myths surrounding the discovery of the heavens and the great forces of nature, which are a part of the collective memory of humankind.

Whenever one enters the domain of mythology one is first impressed and then put off by the often overwhelming array of creeds, rites, and traditional accounts. Folkloric beliefs have a diversity that discourages any attempt at harmonizing them; symbolic systems appear obscure to us today, and we may ask whether they were ever totally transparent. Yet underneath the apparent inconsistencies and contradictions, one occasionally uncovers a faint but unifying link—images and symbols whose related origins can be traced, for example—to help explain those contradictions. Although at the heart of myths there lies a hard nucleus that cannot be subjected to rational analysis,

From the end of the 15th century to the middle of the 19th, yearly almanacs, containing some astronomical information—essentially forecasts of eclipses—and a great deal of astrology, were enormously popular in Europe. Almanacs from Liège in Belgium were especially famous.

In the center of this wheel of the winds, taken from an illuminated manuscript of the 15th century, is Earth, surrounded by ocean and by the cosmic symbol of snakes biting their tails; on the outside edge are the angels of the twelve winds that blow over the land.

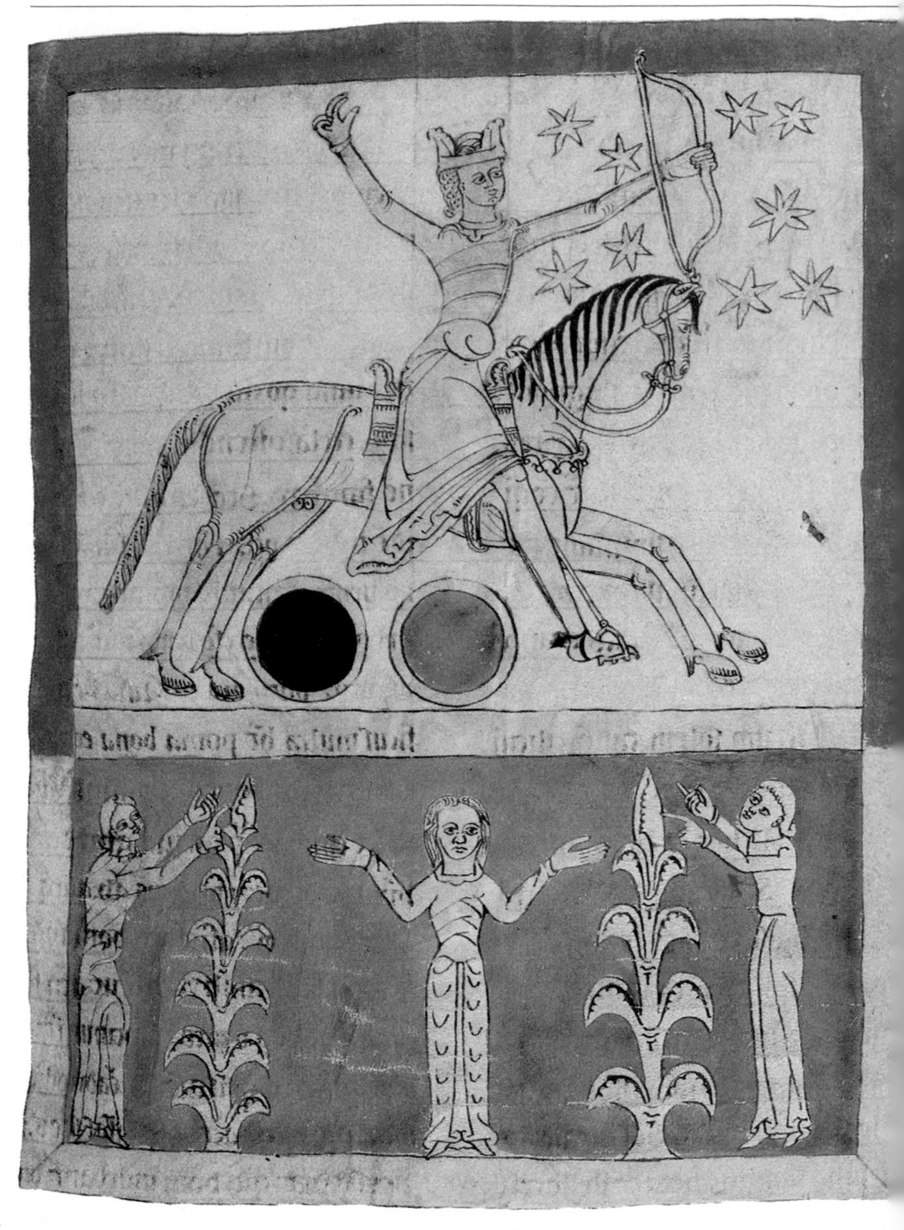

the images that inspired one myth or another and made them perceptible to peoples of the past still speak to our imagination today.

It Is by Means of the Images They Transmit That Myths Have Endured Through the Ages

Myths have often survived the attacks to which they were subjected by means of the imagery through which they are expressed and by virture of their adaptability. Early Christianity in particular has often been cited as a major force in the undermining of many ancient myths, relegating beliefs once held sacred by large groups of people to the realm of folklore. The Christian church naturally found itself in opposition to legends, tenets, and rites that it called pagan, and it sought to remove other systems of belief that could challenge its own authority. In AD 538, at a synod at Auxerre, France, the Catholic hierarchy condemned those who made fountains, woods, or stones the objects of cults. Although this action was clearly intended to be negative, in the end it was only partly so. Christianity

Judeo-Christian legends of a forthcoming Apocalypse, in which God will wipe out all evildoers and reward the faithful with life in a heavenly kingdom, inspired many richly symbolic images. This 12th-century miniature (left) shows one of the four horsemen of the Apocalypse setting out to destroy evil.

In this engraving of the 15th century (below), the two visible luminaries (the sun and the moon) and five visible planets are associated with their astrological symbols. The signs of Aquarius and of Capricorn, for example, correspond to Saturn.

In May 1521, during the course of his exile to the castle of the Wartburg in Thuringia, Reformation leader Martin Luther translated the Old and New Testaments of the Bible into German. Luther's Bible, the first book in the German language, was published in 1534. This colored engraving illustrates Genesis 6:17, in which God said, "For behold, I will bring a flood of waters upon the Earth, to destroy all flesh in which is the breath of life from under heaven."

was in fact offering the people an alternate coherent system of belief, one that was religious—rooted in faith—and therefore also mythical. Moreover, Christianity took over and transformed those myths and rituals it could not totally eliminate, enriching them with new images and new symbols. For example, although the pagan cult of fountains was outlawed by the Catholic church, the rich tradition of symbolism and ritual associated with water was carried forward in the form of baptism, one of the most revered of the Christian sacraments.

Images and Symbols Can Span Different Cultures

As Mircea Eliade, the historian of myth, has rightly pointed out, every culture "is a fall into history," and at the same time every culture is limited. This is true of Greek culture, which, whether we realize it or not, has shaped our own and seems to us to represent a universally recognized ideal of perfection. As Eliade

observes, however, "It is not universally valid, either, since it is a historical phenomenon: Attempt to reveal Greek culture to an African or to an Indonesian, for example; the message will not be transmitted by the remarkable Greek style but by the images that the African or the Indonesian will recognize in the statues or masterpieces of classical literature."

According to the beliefs of one of the most primitive peoples of Asia, the Andamans, the supreme god is Puluga. He lives in heaven, his voice is thunder, the wind is his breath, and the hurricane is the sign of his wrath. When his human subjects began to forget about Puluga, he punished them by unleashing a great flood that was survived by only four privileged people.

In Puluga one can see attributes of Zeus, the king of the Greek gods; his actions also correspond to those of the Hindu god and of Jehovah, the god of the Jewish people. Similarly, much of the imagery relating to these legends transcends cultures and styles.

According to ancient Indian legend, it was Manu who was chosen to repopulate Earth after a great flood. While he was making his ablutions, a fish came into Manu's hand to warn him of the flood and to advise him to construct a boat. This same fish would later pull Manu's boat on the swollen waters. In Hinduism the tale was transformed so that the fish became one of the incarnations of Vishnu and was traditionally represented with blue skin.

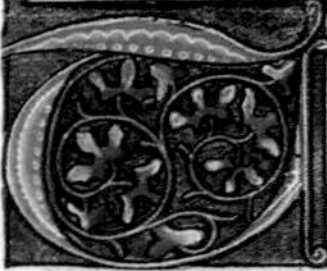

Tout le ciel ne fu onques
fait et ne peut estre
corrumpu si comme au-
cuns dient que si pe-
ut ce estoit plato et ses disciples qui di-
soit que il fu fait et que il est corrupti-
ble mez il est pardurable et a ce prouver
il met .vi. moiens ou raisons car il ne
eut onques commencement ne fin de toute
sa duracion qui est pardurable car il a et
contient en soy temps infini car selo
aristote tout le temps qui fu et sera est
sans commencement et sans fin si comme il
appert en le viii. de phisique et le ciel par
son mouvement contient tout le temps
aussi come la cause contient son effect.

et si comme la mesure contient la chose mesuree et doncques le ciel e sans
commencement et sans fin ce est la pre-
miere raison et selon la translacion
de boece aristote conclut ainsi semper
est eternum sine principio et sine fine
per omnia secula. Il veult dire que le ciel
a dure et durera par tous les siecles.
Cest assavoir que cest mot secu-
lum siecle est prins en .iiii. manieres
Une est pour le monde item plato ap-
pelloit siecle la duracion qui estoit
avant que le commencement du monde
et du temps et ce est eternite pardurable
si comme il fu dit ou .xxviii. chapitre du premier. Item il est dit de la duracion.

The sky proper, that is, the vast space that encompasses the universe and in which the stars and planets exist, seemed to hold little fascination for peasants or sailors. The sun and other stars, the visible planets and Earth's moon, comets, and even meteors gave rise to a large amount of literature, much of it involving superstitions and diverse traditional customs. Yet popular interest did not seem to extend to the celestial vault itself.

CHAPTER II
THE CELESTIAL VAULT

The layers of the sky, as represented in 1377 by Nicholas Oresme in the *Book of Heaven and Earth* (far left). Left: Shou, the Egyptian god of the atmosphere, raised up his daughter, Nout, goddess of the sky, to separate her from the Earth, thus creating the world.

The Sky: A Canopy Spread Over the Earth...or the Shell of a Tortoise?

The sky was often considered to be an immense vault, made of a solid substance, to which the stars were fixed, as in the case of the stars that adorned the vaulted ceilings of churches. According to some beliefs the celestial vault, or the dome of heaven, was made of a liquid material that was prevented from flowing by a high atmospheric pressure and upon which the stars glided like ships on a tranquil sea. This image harks back to ancient western legends of birds that knew the way to the higher ocean. There are more distant links between this image and the legends of the Islamic world, Japanese or Malaysian mythology, and the tales of the heavenly waters that Jehovah released onto a guilty world, in which only Noah and his family were to find his favor.

At one time wrongly identified as a popular engraving from the 15th century, this colored woodcut is in fact a montage created by French astronomer Camille Flammarion for his book *Popular Astronomy*, published in 1880. It represents Flammarion's (and his contemporaries') curiosity about what might lie beyond the outermost layer of the world.

The earthly model associated with this simple representation varied according to the culture: It could be a canopy, cupola, bell, upturned cup, umbrella or parasol rotating around its handle, tent, or tortoise. According to the Gauls, who seemed to consider it not the home of the gods but merely a cradle for atmospheric phenomena, the sky was a solid roof. Their only fear was that one day it might come crashing down onto their heads, yet even if that were to happen, they imagined that they could prop it up with their spears.

According to the disputed interpretation of an ancient ritual text, Pi disks were symbols of the sky in ancient China. The central hole cannot be wider than one third of the disk's diameter. The circle shape is thought to have symbolized the sky in relation to Earth, which was represented as a square.

One Must Go a Long Way Back in the Memory of Humankind to See the Gods of the Sky Seated at the Summits of the Pantheons

Mircea Eliade believes that in most cultures the heavenly deities gradually faded away in the face of other sacred forces that were closer to everyday experience, and therefore more useful, and which eventually came to play the dominant role in determining traditions. Thus, in the Greek history of the sky, Uranus, the old god of the celestial vault proper, was mutilated and dethroned by the youngest of his sons, Cronus. The latter was punished for this patricide by the uprising of his own children against him. At the end of a war lasting for ten years, Cronus and his siblings, the Titans, were defeated and locked up in the sinister darkness of Tartarus in the deepest reaches of Hades.

Supreme authority fell to Zeus, who reigned over the heavens by virtue of his splendor and power, forcing old Uranus, his mythical grandfather, into the shadows. In the case of Zeus personification would go much further than it had with Uranus, whose name had simply designated the sky. Qualifications and attributes became attached to Zeus' name, associating him with atmospheric phenomena: the clouds, rain, thunder, and lightning. Ceremonies and sacrifices were held to appease him. Although in the historical age Zeus was celebrated in art and literature, among the people his influence waned; he became the highest civic god, the protector of political freedom, and festivals were held in his honor.

The Night Sky Presented Patterns of Stars That Every Civilization Arranged into Constellations

Every night the sky is studded with billions of stars that are visible to the naked eye. They have been described variously as tiny openings through which the celestial fire may be glimpsed, as diamonds set in the heavens, or small lakes glittering in the dark grass of the night, as the Eskimos believe, or bubbles of incandescent hydrogen slowly being transformed into helium, as astronomers now know.

Those stars that are high enough above the horizon return every night and assume a seemingly identical position in the heavens. Without ever rising or setting, they revolve around a fixed point in the sky: For those of us who live in the northern hemisphere, this point is the polestar, also called the North Star; for those who live in the southern hemisphere, however, the starry sky turns around a dark area without a visible star. Other stars, closer to the horizon, rise in the East, move across the sky, where they cover a large arc of a circle, set in

As the father of the gods and of humanity, Zeus sits enthroned among his attributes of sovereignty: the eagle, the thunderbolt, and the victory figure. Pictured at left is a reconstruction of the monumental Zeus that once sat in the Greek temple at Olympia, one of the seven Wonders of the World.

This colored lithograph from a late-19th-century catechism illustrates the six days of the Judeo-Christian creation of the world: On the first day God separated the light and the darkness, and on the second day he separated the waters from above and the waters from below and created the firmament. On the third day he created vegetation; on the fourth he created the two luminaries; on the fifth, living beings; and on the sixth, man and woman.

ART 1er JE CROIS EN DIEU
LE PÈRE TOUT PUISSANT
CRÉATEUR DU CIEL
ET DE LA TERRE
IEU CRÉE LA LUMIÈRE
ET SÉPARE LES TÉNÈBR
DIEU SÉPARE LES EAUX DU CIEL DES EAUX DE LA TERRE
DIEU CRÉE LES HERBES ET LES ARBRES
DIEU CRÉE LE SOLEIL LA LUNE ET LES ÉTOILES
DIEU CRÉE LES POISSONS ET LES OISEAUX
DIEU CRÉE LES ANIMAUX DE LA TERRE ET IL FAIT L'HOMME A SON IMAGE

the West, and complete their circular course below the horizon. These are only visible at certain times of the year, and their periodic disappearance does not signify their extinction but quite simply the fact that during this period they cross our sky in the daytime, when the sun floods it with light and when even the brightest stars appear to fade to the point of disappearing altogether.

In order to recognize certain stars more effectively from one night to the next, people quickly got into the habit of grouping them into constellations, associating the brightest of them in geometrical designs of various sizes. The system is so practical that astronomers today continue to use eighty-eight of the ancient constellations (though sometimes with different boundaries) as reference areas in their studies of the sky.

The exact makeup of constellations is arbitrary; every civilization, or every tribe, constructed their outlines, named them, animated them, and dramatized them according to local custom. There exist, however, certain large, characteristically recognizable figures, as well as a few small, very clear groupings, which are all so striking that one finds legends surrounding them wherever they are visible in the sky. These constellations include the hunter Orion, easily spotted by the three bright stars that make up his belt; Taurus the Bull, which incorporates the asterism (a distinctive star group within a larger constellation) known as the Pleiades; Gemini, marked by the twin stars Castor and Pollux; and the famous Ursa Major, the Great Bear, which contains another asterism, the Big Dipper.

The images people have placed in the sky reveal their interests and preoccupations. At a time when humans hunted to survive, they saw dogs, bears, and hunters in the sky, as illustrated in this 14th-century astrological treatise. In the 18th century European navigators arriving in the southern hemisphere placed telescopes, microscopes, compasses, and ships' sterns in the sky.

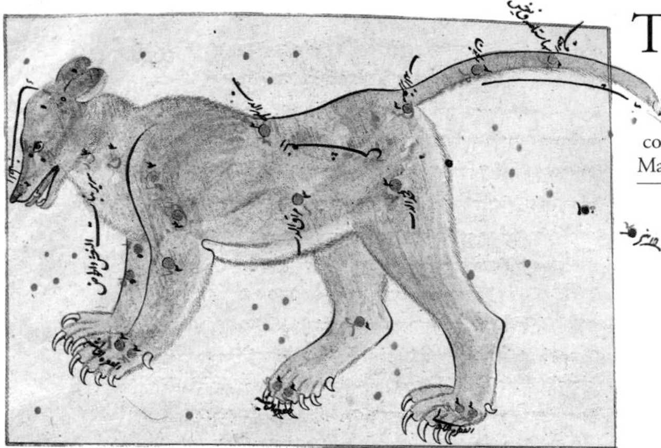

This 15th-century Persian manuscript presents a surprisingly gentle image of the constellation Ursa Major, the Great Bear.

The Great Bear, Often Called the Heavenly Chariot, Was Named After a Greek Legend

According to the legend, Callisto, thought to be either a nymph or the daughter of a king, was loved by Zeus, and she became pregnant by the king of the gods. Some versions of the legend say that it was Zeus' wife, Hera, who transformed Callisto into a female bear as punishment; others maintain that it was a scheme devised by Zeus in order to hide his lover and thus shield her from his wife's jealousy. In either case it was Zeus who projected Callisto into the sky and made her into the constellation of the Great Bear in a supreme act of glorification.

According to the Aztecs this constellation represents Tezcatlipoca, a dark god associated with death and the North. The god is missing one foot, which was thought to have been devoured by a heavenly monster. In fact this aspect of the Aztec legend relates to the visibility of the constellation from different regions of the world. At northern latitudes the Great Bear never sets, but at the southern latitudes of the Mexican plateaus on which the Aztecs lived, the last star of the Great Bear disappears below the horizon.

In Hindu mythology the seven brightest stars of the Great Bear represent the homes of the seven Rishi, or the seven primordial sages. Similarly, the Chinese consider these seven stars to be images of the seven astronomical Rectors, the masters of the reality of heavenly influences; however, they also see them as the seven openings of the heart.

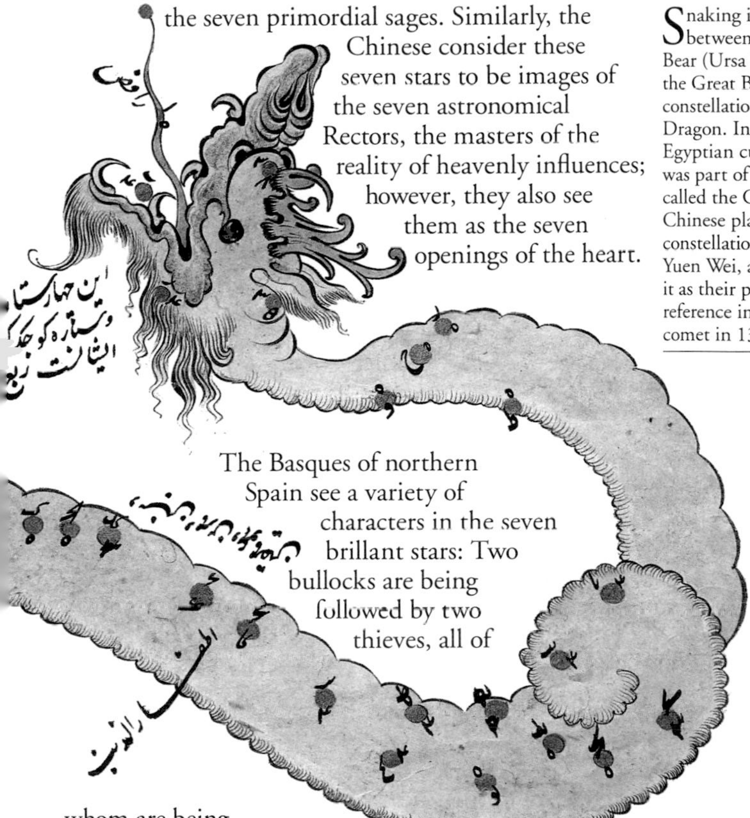

The Basques of northern Spain see a variety of characters in the seven brillant stars: Two bullocks are being followed by two thieves, all of whom are being watched by a herdsman and his male and female servants. In this version of the tale there is an eighth star, fainter than the others, called Alcor, which represents a small dog. In variations on the legend, Alcor, which can be found above the second bullock, is a small rat that gnaws at the straps of the bullock's yoke.

Some legends interpret the Great Bear as a saucepan. In those stories Alcor is a small man who waits for the moment at which the saucepan's contents will begin to boil in order to take it away from the fire. That day will be the last day of the world.

Snaking its way between the Little Bear (Ursa Minor) and the Great Bear is the constellation Draco, the Dragon. In early Egyptian culture Draco was part of a constellation called the Crocodile; the Chinese placed it in a constellation they called Yuen Wei, and they used it as their point of reference in observing a comet in 1337.

The North Star Is the Landmark of All Wanderers of the Earth, Sea, and Sky

For some people the stars were the windows of the world and, for others, its eyes, from which burst rays of light or from which insects pour forth toward Earth. For certain civilizations in the northern hemisphere the North Star was seen as the opening in the sky that linked different worlds and through which heroes passed on their way to join the gods or to return from the heavens to Earth. The position of other stars can be determined by this star, which contributed to the idea of the stars as horses and the North Star the stake to which they are attached. It is by means of this star that nomads, navigators, and, later, the first aviators, got their bearings.

In this engraving, taken from *The Great Calendar and Compost of the Shepherds,* one of the most famous of the early almanacs, a shepherd calculates the hour of the night from the passage of the stars in relation to a vertical line starting at the North Star.

The drawing below depicts the sky and the lower worlds as seen by the Chukchi people of eastern Siberia. The tiny group of seven stars at top center are the Pleiades, and the Milky Way runs between two parallel lines on the lower left.

Venus Sits in Majesty As the Queen of Planetary Mythology

One half painted red, the other black, this Iroquois mask symbolizes the East and the West and the morning and the evening.

"The morning star looks like a man who is completely covered in red paint; that is the color of life. He wears leg warmers and a cloak. On his head lies a soft and downy eagle's feather, painted red. This feather represents the soft and light cloud that floats high in the sky.... Morning Star, give us strength and renewal.... The day is on its heels." So went a chant of native tribes living on the prairies of North America. Paradoxically, the morning star, which others call the shepherd's star, is not a star at all but a planet, the planet Venus.

As a planet Venus does not glitter in the manner of the stars, which dispose of enormous stores of nuclear energy at their cores. Venus shines only by means of the light that reaches it from the sun, which is reflected by the thick clouds of its atmosphere. In the bluish white of dawn, Venus gives off a brilliant white light. That the early North Americans saw it as redder than Mars could reflect a vivid imagination or the possibility that their eastern horizon was saturated with dust that reddened the solar light. And the line of the song referring to the day following at Venus' heels indicates that they quite rightly noticed that Venus,

The plumed Serpent Quetzalcoatl is the most important of the Aztec gods. In his various forms, he embodies the sun, the wind, and the planet Venus.

which revolves between Earth and the sun, never moves far from its source of light.

Charming or Pernicious, Venus Is Never Regarded with Indifference

In spite of its brilliance, which ought to imbue it with all the positive images that are associated with light, and even though it accompanies the dawn with its wealth of possibilities, Venus is not always considered a benevolent planet. The ancient Mexicans feared it and locked their doors and windows before dawn to protect themselves from its rays, which they imagined carried

Atlantes are male figures used as columns to support the roof of a building. This figure, from the Mexican site of Tula, represents Quetzalcoatl in the form of Venus.

dreadful diseases. The Maya saw Venus as the sun's elder brother, a corpulent man whose massive face was adorned with a long beard. As in Hindu legend, the idea of Venus as the brother of the sun can be related to the fact that Venus always remains close to the sun, rising before it and following it as it sets. It may be this same phenomenon that explains its maleficent aspects: It is seen only briefly, first in the East, the direction of the day, then in the West, the direction of the night. For the Maya and the Aztecs, Venus became a symbol of death as well as rebirth in the form of the Aztec god Quetzalcoatl, by whom the human race was first wiped out and then re-created from some bones that were

Pietro Vannucci (c. 1448–1523), known as Il Perugino, painted Venus standing on her chariot, which is being drawn by two doves. She is accompanied by her son, Cupid.

removed from the realm of the dead and brought back to life by the blood sacrifice offered by this god.

Western tradition, however, associates Venus with tenderness, pleasure, and love. This was especially true in ancient Rome, where the old Italic deity Venus was identified with Aphrodite, the Greek goddess of sensual pleasure and physical love. Venus was also made one of the protectors of Rome, and in the Middle Ages, was known as the "little beneficent one."

The zodiacal chart below comes from a 15th-century Italian manuscript called *De Universo.*

Like Venus, the Other Planets Visible from Earth with the Naked Eye Received Their Names, and Their Attributes in Western Culture, from Roman Mythology

The planet Mars was at one time called the "little evil one"; its namesake, the Roman god of war, signifies energy, ardor, and aggression. The offspring of the sun and the moon, Mercury holds a privileged position as a mediator, representative of the principles of linkage, exchange, and communication. Jupiter, the largest of the planets, enjoys powers befitting his rank. The Roman counterpart of Zeus, Jupiter embodies the principles of authority, order, and balance. And Saturn, who as the father of Jupiter was equated with the evil Cronus of Greek myth, imparts negative connotations: he is the "great evil one," symbol of impotence, misfortune, and paralysis.

Since humans were thought to represent a reduced version of the cosmos, many astrological maps, such as the one at right, from *Les Très Riches Heures du Duc de Berry,* feature the human body. In the center of the map each of the twelve signs of the zodiac is placed at the part of the body it is thought to govern.

Since Ancient Times Astrologers Have Connected the Movement of the Planets to Daily Life on Earth

In charting these movements, astrologers look to the ecliptic, the apparent path of the sun across the sky during a calendar year. In the course of their orbits around the sun, the planets all revolve within a narrow band, which incorporates the ecliptic at its center. Following this band, the planets and the sun and the moon cross twelve constellations that make up the signs

of the zodiac: Aries,
Taurus, Gemini, Cancer,
Leo, Virgo, Libra, Scorpio, Sagittarius, Capricorn,
Aquarius, and Pisces.

The zodiac owes its name to the bestiary that inhabits it. The twelve signs were also called "houses of heaven" or "monthly residences of Apollo," referring to

Aries 21 March–19 April

Taurus 20 April–20 May

Gemini 21 May–21 June

Cancer 22 June–22 July

Leo 23 July–22 August

Virgo 23 August–22 September

Libra 23 September–23 October

Scorpio 24 October–21 November

Capricorn 22 December–19 January

Sagittarius 22 November–21 December

Aquarius 20 January–18 February

Pisces 19 February–20 March

the Greek and Roman god of the sun who visits a new "house" every month and returns every spring to begin again.

It should be noted, however, that when astronomers speak of Aries, they are referring to the *constellation* Aries, whereas when astrologers speak of Aries, they mean the *sign* of Aries, that is, one of the twelve areas of the zodiac that the sun takes a month to cross. There was a time, in the 2nd century BC, when signs and constellations coincided with each other; unfortunately, the precession of the equinoxes, that slippage of the

More than two thousand years ago astrologers invented a cartography of fate, represented by the horoscope. By diagraming the relative positions of the planets and the constellations of the zodiac at specific times (at birth, for example), astrologers believe they can evaluate a subject's personality and predict future events. Meetings with astrologers (left) and the casting of a newborn baby's horoscope (below) are echoed in today's abundance of available astrological literature.

The lack of free will implicit in astrology offended the 14th-century Italian poet Petrarch, who believed that human beings were masters of their own fate: "Why do you degrade the sky and the earth and pointlessly humiliate children? Why do you impose your futile laws on the brilliant skies? Why make us the slaves of an inanimate sky, we who have been born free?"

sun's "path" through the stars, has played a nasty trick on astrologers: As the point at which the sun crosses the equator in the spring is progressively altered, so are the astrologers' signs. The effect of this is that today an astrological sign falls across a constellation different from the one that gave it its name. For example, the sign of Aries now covers the constellation of Pisces; however, astrologers believe that the area of the sky in which the constellation of Aries used to lie, more than two thousand years ago, has retained that animal's positive virtues (Leo continues to represent strength, Gemini tenderness, and so on). Still, the dates applica-

This 15th-century miniature from the *Book of the Property of Things,* by Bartholomaeus Anglicus, represents the astrologer's sky, with personifications of the five visible planets and two luminaries surrounded by the signs of the zodiac. Pages 44–7: These 15th-century Italian engravings depict, respectively, the sun, the moon, Saturn, and Jupiter.

• SOL •

• LVNA •

·SATVRNVS·

• JOVIS •

ble to each sign are only approximate and can vary slightly from year to year. This is why a person whose birthday falls on the cusp of two signs is often thought to possess a combination of characteristics pertaining to both signs.

A Milky Trail in the Heavens, the Milky Way Galaxy Is the Path Souls Follow to the Land of the Dead

On a very clear night the attentive eye can make out a long, whitish ribbon, shining faintly across a wide stretch of sky, while elsewhere the sky's black background is peppered with very bright stars. For a long time the Milky Way remained a mystery that early astronomers explained as exhalations given off by Earth as it simply floated in space. It was Galileo who pointed his telescope at the milky patch of sky and discovered within it a great multitude of stars.

European tradition saw this whitish cloud as a smear of milk stretching across the sky, and the origin of its name derives from a legend involving Hercules, one of the most popular heroes of Greek mythology. In order to attain immortality, the baby Hercules had to take milk from the breast of Zeus' cantankerous wife, Hera. Hermes, Zeus' gifted son (and the inventor of the flute), placed the child on the breast of the sleeping goddess. As soon as she opened her eyes, Hera pushed the young Hercules aside, but it was too late. The milk that had begun to flow from her breast formed a trail in the sky that became the Milky Way.

For the Eskimos the Milky Way represents the snowy path of the Great Raven. In Estonia and Lapland the Milky Way is the path of migrating birds, and for the Finns of the Volga, the path of wild geese in particular. For the Tartars of the Caucasus it marks the flight of a thief who trails part of the straw he has stolen behind him. To some Muslims the Milky Way marks the path of pilgrims on their way to Mecca, while for European Catholics it is often seen as the pilgrims' path to the cathedral of Santiago de Compostela in Spain.

One African tribe in Botswana thinks of the Milky Way as the spine of the night, as if the night were a huge animal inside which we live. They believe the Milky Way holds up the night, and without it fragments of darkness would crash at our feet.

Legend further states that Saint James appeared to Charlemagne in the Milky Way in order to show the Carolingian ruler the pathway to Spain and the place in which the saint's tomb could be discovered. All over the world the Milky Way is considered the path of souls departing for the afterlife: At its end is the land inhabited by the dead.

In numerous traditions the Milky Way appears as one of the nocturnal passageways linking Earth and Heaven, a role that is filled during the daytime by the rainbow. The Chinese, for whom the Milky Way is the great heavenly river, have a similar view: This celestial river leads to the bottomless abyss of the southeast, in which the mother of the sun and the mother of the moon bathe their children every morning before they take their places in the sky.

In 1519, almost thirty years after the arrival of Italian explorer Christopher Columbus in the West Indies on behalf of Queen Isabella of Spain, the Spaniard Hernán Cortés conquered Mexico and encountered the peoples of the sun. As Moctezuma, the emperor of the Aztecs, showed the conquistador to the top of the pyramid that overlooks what is now Mexico City, Cortés would have seen the figures of dragons adorning the stones, as well as large amounts of freshly spilled blood from the human sacrifices that were made there.

CHAPTER III

STAR OF DAY, STAR OF NIGHT

Michael Wolgemut's engraving of 1491 (left) shows Christ walking with Death, with the sun and the moon looking on. Right: A scene from a facsimile of an Aztec codex, or historical manuscript, depicting a ritual sacrifice.

The Sun's Chosen People, the Aztecs, Ensured the Survival of Their God by Means of "Precious Water," the Blood of Sacrifices

Many of the ancient peoples of Mesoamerica believed that prior to our present universe there had existed four worlds, and all had been destroyed by cataclysms. A fifth world, our own, was formed out of the combined actions of two opposing gods: Quetzalcoatl, the plumed serpent, god of light, and Tezcatlipoca, the smoking mirror, god of darkness, who had lifted up the sky after the last catastrophe. But this newly created world was dark and cold, and the other gods in the pantheon decided to create the sun and the moon, which were to be born of the sacrifice of two of their number. Quetzalcoatl jumped spontaneously into the sacrificial furnace and reemerged instantly as the sun. Tezcatlipoca hesitated four times before making his jump, and that is why the moon is less bright than the sun.

The world did not yet function properly, however. The stars, which were motionless in the sky, were in danger of setting Earth on fire. In order to set the stars in motion, a sacrifice of all the gods had to be made. Quetzalcoatl had the responsibility of massacring his fellow gods before committing suicide. But once again after sacrificing himself Quetzalcoatl came back to life, whereupon he went down into the underworld to look for the bones of the dead gods. He created a new human race by crushing the gods' bones to a powder and soaking it in his own blood. Hence, a great deal of blood and many deaths were necessary in the creation of the world, and from this came the notion of death as the indispensable source of life.

The Aztecs drew the following lesson from this myth: The world, and therefore human life, depends on the movement of the sun; the world is fragile, however, as is its sun. It was the emperor's responsibility to ensure that the world worked well, that the pact linking the

The Aztecs made many human sacrifices as part of elaborate ceremonies dedicated to the sun or to other great divinities. Cortés fought to suppress them.

world with the gods who had created it was upheld, and that the "precious water," the blood of sacrificial victims, flowed on the altar and commanded the respect of the forces of destruction.

This impressive monolith weighing twenty tons, the "stone of the sun," is a record of the cosmology of the Aztecs. In the center of the circle is a human face with its tongue hanging out, traditionally interpreted as that of Tonatiuh, the sun god, who demanded offerings of human blood.

In Babylon Homage Was Paid Above All to the Moon-God, the Supreme Guarantor of Cosmic Order

Seventeen centuries before Christ Babylonian civilization produced the first literary work recorded in the history of humanity—an account of the creation of the world known as *Enouma Elish:* "At the beginning,

The scene on this carved stela represents the Babylonian king Melishipak II presenting his daughter to the goddess Nanai.

the gods Anou, Enlil, and Ea divided everything up between the two gods who were the keepers of the sky and the earth.... Sin and Shamash were given two equal portions, day and night."

The sun and the moon were born together from the body of the mother-goddess, Tiamat. Masters of the light of day and of the light of night, respectively, Sin and Shamash also are the masters of time: The sun, which gives us both the days and years, is a master twice over, and the moon, which separates the months, is a master only once.

For Babylonians the moon, which waxes, wanes, disappears, and is reborn, was the ideal symbol of becoming and of precariousness; hence, the fragility of the moon had to be protected. If every new moon (when the moon is not visible at all) represented the temporary death of Earth, the eclipses, which brought with them the risk of permanent death, were even greater objects of fear. In order to ward off the evil effects of eclipses specific rituals were required, depending on the months in which the eclipses

This colossal head symbolizes Coyolxauhqui, Aztec goddess of the moon and of the night; she bears the phases of the moon on her forehead. Coyolxauhqui was the sister of Huitzilopochtli, the god of war and symbol of the sun at its zenith.

occurred: One had either to wash the king with turpentine and cover him with oil of myrrh or to lay him down behind a door and sprinkle him with rainwater before dressing him in festive clothes and letting him kiss an old woman. As was true in Mesoamerica, in Babylonia, also, the temporal ruler was responsible for the smooth running of the world.

The Sun and Moon Are Closely Related in Ancient Myth

The sun and the moon are often considered opposites, each ruling a different portion of the sky. One controls the day, the other, the night. At the time of the full moon, when the two bodies are aligned on either side of the Earth, the moon rises in the East at the very moment the sun disappears in the West.

In many civilizations the horns of bovine animals were used as symbols of the moon because of their shape, which recalls the lunar crescent. The Assyrians linked the moon with a Great Cow in their fertility rites, and a Sumerian hymn refers to the moon god as an "alert bull with untiring hooves." Above is a bit from a horse's bridle, found in western Iran.

According to Babylonian and Mesoamerican mythology the sun and the moon were born together. Very often in cosmogonical accounts they are even related to one another. The Eskimos, for example, believe that the sun and the moon are two children from a coastal village. The girl, who is being bothered

Alchemists were medieval scientists who, through the use of chemicals, sought a universal cure for diseases and a means of changing base metals into gold. This image of the sun and moon in majesty comes from an alchemical treatise of the 16th century. The moon is dressed in white while the sun wears red; the Earth catches fire beneath the sun's feet, a reflection of alchemists' belief in the sun as the innate fire in matter. They also called the moon Diana and the sun Apollo: In Greek mythology Diana is Apollo's elder sister, and she acted as her mother's midwife in order to bring her brother into the world; thus the color red, which was taken by the sun, appeared later than white, which is called moon. This is why in alchemy red sulfur was given the name solar tree and white sulfur was referred to as lunar tree.

by her brother, escapes, climbs up a long ladder, and becomes the sun. Her brother, without even stopping to get dressed, rushes off in pursuit of her and becomes the moon, which will never catch up with the sun. Eventually, weak from hunger, the boy-moon faints; in order to keep him alive but still unable to catch up, the girl-sun continually feeds the boy-moon and then deprives him of food until he faints again. Interestingly, in this explanation of the phases of the moon the Eskimo breaks with the traditional association of the moon, the dark and cold star of the night, as a woman.

An anonymous 19th-century engraving illustrates a popular peasant folktale of the time: "Old man, you have once again come to steal wood from us." "Faith, may I be in the moon if I am lying." He had not even finished speaking when he found himself there.

In our own time the belief in a dualistic creation of the world has not disappeared from peasant folklore. In Brittany lists are drawn up of the works of God and of the Devil. If God created the sun, the Devil must have created the moon, since the moon is very often considered a failed or fallen sun. In the south of France a slightly different version holds that God had two suns and was keeping one of them in reserve; one day, not knowing what to do with the one that was growing old in a corner, he threw it at the sky and made it into the moon.

The Man in the Moon

The moon always shows us the same dull face, enlivened only by patches or spots of varying size. Many legends trace these spots to human or animal origins. They represent a man who, in order to atone for a wrongdoing, has been transported to the moon where he may be seen nailed to the celestial pillory as an example to all. In order to make the lesson even more striking, the man bears on his back the evidence of his crime, which most often is a religious one; in Christian countries it generally involves a violation of the Sunday rest, a theft, or a failing in charity. The evidence is usually a bundle of firewood, which is connected to crime in that the guilty person either cut it on a Sunday, stole it, or refused to give it to someone who was poorer than he.

In Gascony it is the wind that each Easter Sunday

sends a habitual offender off to the moon, with the wood he had cut in order to repair his hedge, to await Judgment Day. At the beginning of our century the peasants of the groves and fields of the Vendée showed their children the man who had refused to admit Jesus to the warmth of his hearth and who had been banished to the moon with his bunch of sticks. In western Brittany the people believe that one day God appeared to a man who had just stolen a load of wood, saying, "This wood does not belong to you and in order to punish you I should let you die, but you shall die when your time comes; afterward, I shall let you choose where you are to be exiled: either to the sun or the moon." "I prefer the moon," replied the thief. "It only walks at night and I shall not be seen as frequently."

The presence of firewood fits well with legends in which the man in the moon is punished for theft, the stealing of wood being a crime that was usually committed at night. In many countries, in fact, stolen wood is called the wood of the moon. Its presence in other cases is also interesting. Some folklorists trace its origin to the Bible where, in the Book of Numbers, Jehovah asks Moses to put to death a man who had been caught gathering wood on the Sabbath.

Shinto, the native religion of Japan, gives little place to the moon and its god, thus leaving the field open to other popular beliefs. The image of the hare as an inhabitant of the moon came to Japan, and to China, from India with Buddhism. The Japanese quickly assimilated it as the white hare of Inaba, whose story is told in the *Kojiki,* a collection of legends preserved in an oral form, which was completed in AD 712. The illustration at right comes from the Japanese folktale *The Hundred Beauties of the Moon.*

Most of the Mexican codices show the moon as a crescent-shaped vessel full of water in which the outline of a rabbit appears.

Since the Beginning of Time Human Imagination Has Populated the Sun and the Moon

"I have seen three small rabbits on the moon...." This nursery rhyme recalls a series of legends from various countries that show the moon inhabited by a rabbit or a hare. The Hottentots of southern Africa associate the hare with the moon in the following legend:

One day the moon ordered a louse to announce to all the humans that their fate would be similar to that of lice, which die only to be reborn. On its way the louse met a hare who declared that he could carry the message to the people faster. But hares lose their memory as they run and this one forgot part of the message; he simply told people that, like the moon, they too would decline and die. The moon was

月百姿
玉兎
孫悟空

extremely displeased at the fact that his message had been distorted. He brandished a piece of wood and struck the animal on the lip, and ever since that time the hare has had a split lip.

It is much less common to hear of tales in which the sun is inhabited. In the west African country of Togo, however, the Dagombas say that on the sun is a field for festivals, which may be seen clearly when the sun is surrounded by a halo. In this field lives the ram of God. Whenever the animal strikes the sun with its hooves, thunder roars, and whenever it shakes its tail lightning flashes. Rain is caused by tufts of wool falling from the ram's fleece, and when the wind blows, it is because the ram is galloping around the field.

The Moon and the Sun Mark Out Time

Even though the moon is often considered a fallen sun, and even though it has fewer effects on Earth than the sun does, its influence among earlier peoples was strong, and numerous powers were attributed to it. Many of these powers were merely superstitions, such as the old wives' tale that a woman who urinates facing the moon in the evening will become pregnant.

The moon's primary power, however, derives from observed fact. As a guardian of time, the moon marks out the twelve months of the year. As the other guardian of time, the sun is responsible for the alternation of day and night and for providing the rhythm, or seasons, of the years. Yet the sun always appears in its same form, seemingly unchanging; the moon's phases mirror the progression of human life, where time passes, people grow older and pass away, and their offspring continue the cycle.

The moon, then, offers us a concrete, living time that passes, a time that we understand easily. As the mistress of time, the moon can also be seen as a mistress of fate. A Babylonian tradition asserts that man was created at the time of the full moon. The fact that creation and rebirth coincide guarantees that the created beings will continue to grow in sync with the moon.

This 17th-century French engraving satirizes the notion of the moon as having an influence on women.

Although some cultures believe in the concept of reincarnation, when a soul is reborn in a new human body, humankind is not perpetually reborn the way the moon is. An African tale explains: One night a village elder came upon a dead man upon whom the moonlight was shining. He brought together a large number of animals and asked them, "Which of you wishes to take either the dead man or the moon to the other side of the river?" Two tortoises came forward.

The coincidence between the phases of a lunar month and those of a woman's menstrual cycle encouraged the popular belief of the moon as woman's first husband.

The Four Seasons

From the time of the emperor Charlemagne (9th century) the French year began at Christmas. Christmas Day was doubly important in that it celebrated the birth of Christ and, by virtue of its proximity to the winter solstice (usually around December 21) it was the legal day on which the year was renewed. This 15th-century calendar presents the months in a sequence similar to today's. Each month is evoked by an activity or an atmospheric phenomenon: In January it is snowing, but by February the softening ground can be cleared. Vines are pruned in March, and in April lambs are born. Hunting begins in May, and hay is made in June. July marks the annual harvest and August the threshing of the corn. New seeds are sown in September and grapes pressed in October. In November pigs are led out to be fattened for slaughter in December.

The first, who had long legs, carried the moon safely to the opposite bank; the second, who had short legs, tried to take the dead man across and was drowned. That is why when the moon dies it always reappears, whereas a person who dies does not return.

The Moon as the Mistress of the Waters

Ancient peoples saw the moon as reigning over the waters from above. Some people still believe today that the rain goes away or returns with the new moon; yet when the new moon appears it is seen from all over Earth, and we know that rain does not start or stop everywhere at the same time.

The moon can be said to reign over the waters down below: From very early times people noticed that the sea rises and falls in relation to the rhythm of the moon, and, for once, the appropriateness of the analogy has not been disproved by scientific progress. In fact it is the gravitational pull of the moon, combined with a less powerful attraction from the more distant sun, that regulates the tides, the great oscillating movement of the oceans.

Many tales and legends describe the moon's connection with the waters. In animal-based legends the wolf is often duped by the combined force of the moon and water, and it is often the fox that deceives him. About to be devoured by the wolf, the fox shows the wolf the reflection of the moon in a pool of tranquil water and convinces him that there is a girl bathing in it. Hearing this, the wolf jumps into the water and drowns.

It Is Perhaps in Indian Mythology That the Link Between the Moon and Water Is Most Apparent

It is Soma who maintains the link between Candra, the Hindu lunar god, and the waters. The name Soma derives from a milky, fermented liquid found in a plant growing in the mountains of eastern India; according to the Vedic hymns (which are among the earliest sacred Hindu writings), it constitutes the hair of those mountains. In order to be purified the sap of the plant

The churning of the sea in order to obtain the potion of immortality is one of the most popular motifs in Hindu mythology. Vishnu, incarnated in a tortoise, is the pivot on which Mount Mandara is made to turn. Demons hold one end of the snake, Vasuki, with which they control the turning of the mountain; the gods hold the other end. The tortoise is often seen as a symbol of the cosmos because of its shell, with its rounded top that recalls the celestial dome and its flattened bottom that represents the Hindu conception of Earth. Its compact shape and slow stubborn strength, especially evident in the short but powerful legs, lend credence to its role as the bearer of the world. After the churning of the sea the god Indra distributes the soma (below).

must be extracted and passed through a sieve made of lamb's wool. The soma is then poured into wooden vases and mixed with water and milk.

This bittersweet liquor, which is slightly intoxicating (it unties the tongues of poets), was endowed with all virtues and all powers and rapidly became the nectar

of the gods. These virtues and powers were soon personified under the name of Soma, and Soma attained the rank of a prominent god. The ritual of the extraction of the soma took on a cosmic dimension: The filter that was used symbolized the sky, and the sap itself represented rain; in this way Soma became the lord of the waters. Eventually Soma absorbed Candra, who had appeared while the ocean was being churned with milk in order to draw soma from it. Candra thus became merely another name for the lord of the waters, who every evening emerged once again from the ocean to inhabit the sky.

Nimba is the goddess of fertility for the Baga people of northwestern Africa. Nimba protects pregnant women and cures sterility. Her large breasts remind one of the large number of children she has carried. Her face resembles the calao, a bird traditionally associated with fertility by the Baga people. Nimba is displayed during ceremonies celebrating the rice harvest and is carried by strong young men who are hidden beneath her costume of wood fiber. This statue, acquired in 1933, is one of the most important works in the Musée de l'Homme in Paris.

The Moon, a Celestial Gardener

For those who perceive the moon as mistress of time, reigning over evolution as well as over the germinal waters, the moon serves also as mistress of all vegetation. According to the *Yasht,* an Iranian text, it is by the heat of the moon that plants grow. Among certain tribes of Brazil the moon is considered the mother of grasses, and in ancient China it was believed that grasses grew on the moon. In many places even today peasants often sow their seed at the time of the full moon, thereby guaranteeing for the seed a period of growth that is in sympathy with the waxing of the lunar disk; on the other hand, they prefer to prune trees and harvest crops when the moon is waning, doubtless fearful of going against the cosmic rhythm by breaking a living organism while the moon is still growing.

The traditional gardener's fear of the so-called red moon is well known. This phenomenon begins in April and ends in May, a time when the young shoots are fragile and when morning frosts may still occur. It is neither the moon nor its light that freezes young plants, but the clarity of the moon and the brightness of its light are the signs of an especially transparent sky. If the sky is very clear the soil cools down very quickly as soon as night falls, which can cause the temperature to go quickly down to freezing. If on the contrary the sky is covered with clouds and the moon does not appear, the cloud cover slows down the cooling of the ground, which poses less of a threat to vegetation.

All of these correspondences between the moon, the time that passes, the rain, and vegetation are expressed in the religion of the Pygmies. Among these African peoples the festival of the new moon takes place immediately before the rainy season and is reserved exclusively for women (whereas the festival of the sun is reserved for men). In order to glorify the moon, which is not only the "mother of vegetation" but also the "mother and home of ghosts," the women smear themselves with the sap of plants and with clay in order to become as white as specters and as moonlight. They dance until they are exhausted and drink an alcoholic drink made from fermented bananas. As they drink they cry out and beg the moon, "mother of living things," to ward off the spirits of the dead and to give the tribe many children, fish, game, and fruit.

Throughout History the Sun Has Been Honored as King of Our Universe

Although the sun is an ordinary star, of medium size and temperature, it has the fundamental quality of being close to Earth. The sun is Earth's star. For more than five billion years it has been flooding Earth with light, from

Louis XIV, who ruled France from 1643 to 1715, presented himself as the Sun King, and he commissioned countless portraits to reinforce this image. He is shown here as Apollo.

Eighteenth-century naturalist Johann Jakob Scheuchzer is remembered for his theories about fossils and their relationship to Noah's flood. In his *Biblia Sacra* he compares the Judeo-Christian Scriptures to the religious texts of other civilizations. This engraving from Scheuchzer's bible illustrates Japanese rites in honor of the sun. In ancient Japan the forces of nature were venerated as superior beings called Kamis. An important Japanese ritual celebrated the moment at which the sun reaches its lowest point in the sky during the winter solstice. This ceremony was performed mainly by women in order to renew the sun's declining life force.

which all forms of energy and life were born. In almost every culture the sun has been honored, and often deified, but actual solar cults are much rarer than one might imagine. They are found only in certain areas of the globe. At the beginning of our century the great ethnologist Sir James Frazer noticed "inconsistencies" within the solar elements in the mythologies of Africa, the Pacific Ocean, and Australia. The same lack of unity applies to ancient cultures of North and South America, with two significant exceptions: the Inca empire of Peru and the Aztec empire of Mexico, that is, in the only two American peoples who have in the past developed vast political organizations. This same unifying factor holds true throughout the rest of the world. It is only among the politically organized civilizations of Egypt, ancient Europe, and Asia that the cult of the sun enjoyed such great favor, which confirms one of the great social functions of myths, to justify the political structures of a society: The king or emperor, son of the sun, reigned over the social order while the sun reigned over the cosmic order.

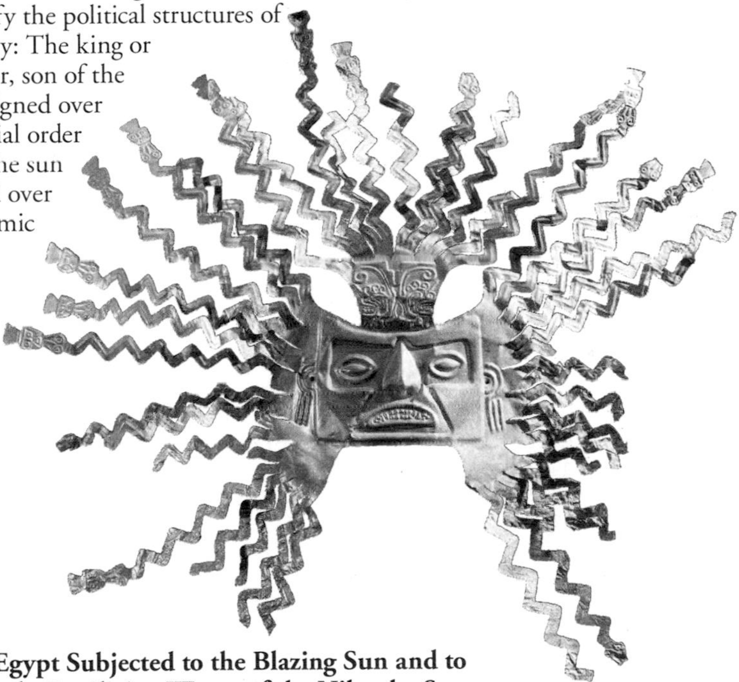

In an Egypt Subjected to the Blazing Sun and to the Black, Fertilizing Waters of the Nile, the Sun-God Represented the Top of the Pantheon

The Egyptians, who preceded the Aztecs in history, formed the most elaborate and glorious of the solar cults. It happened very early on that the sun-god Ra, who was honored at Heliopolis, began to absorb other divinities. The earliest fusion between the gods took place around 3000 BC. At that time the pharaoh Menes adored Horus, the falcon-god, who had the sun as one eye and the moon as the other. Menes created the first

This gold Inca mask represents the sun. A tribe of the Quechua peoples of Peru, the Incas made up a privileged clan that founded a powerful empire with a centralized government. The leader, who was called the Inca, was venerated as the son of the sun.

Hent-Taoui was an Egyptian priestess-musician of the sun-god Ra. Together the priestess and the lunar god Thot, who is depicted with either an ibis or a baboon's head, adore the solar disk, which is adorned with the eye of Ra. It was Thot who revealed to the dead the magic formulas that they needed to enter the lower worlds.

imperial dynasty and founded the capital of a unified Egypt at Thinis, near Abydos, the city that was sacred to Osiris, the god of fecundity. He later constructed the imperial city of Memphis at a site not far from Heliopolis, the great center of the solar cult. Mythology eventually combined Horus, Osiris, and Ra, and the idea of the sun-god became so inseparable from that of the god of the state that all the local divinities eventually assumed a solar aspect and enriched the solar god with their own qualities.

Accompanied by the bennu, an imaginary bird born of fire, the falcon-god Horus represents the sun as it appeared at the moment of creation.

The Egyptian Sun-God Reigned in Many Different Forms

The sun, having thus become the most important of all the Egyptian divinities, bore several names. As the solar disk itself he was called Aton. But as the rising sun he became Khepri, in the guise of a giant scarab, pushing the solar globe before him just as the scarab pushes a little ball of dung, a reserve of food for the darkest days, in which the Egyptians believed the scarab hid its eggs, out of which life would reemerge. At its zenith in the sky the sun was called Ra, the god of Heliopolis. Finally, as it set it became the old man Atoum. The sun-god also assumed the name Horus and, when the people wished to combine the qualities of Ra and Horus, it was called Ra-Horakhti. Assuming the form of a winged disk, Ra-Horakhti emerged on the horizon in a splendor that was renewed every day.

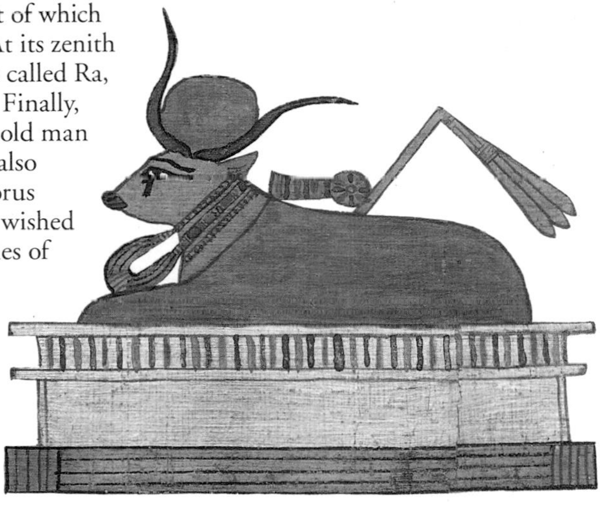

Other beliefs held that every morning a celestial cow gave birth to this golden calf, the sun, and that every evening the woman of the sky opened her mouth to swallow him, or that the sun was an egg laid every morning by the "Great Cackler," Geb, the god of the Earth, who took the form of a moorhen. But the most familiar image of the sun's progress across the sky is that of a barge. To sail across the sky Ra had at his disposal two vessels: for the daytime, the "boat of millions of years," and, for the night, the "boat of darkness," of the dead.

The bull Apis represented one form of the god of the Nile. He was also considered to be an incarnation of the son of Osiris. In his association with the solar cult (he was thought to have been conceived when a ray of light fell onto a cow), Apis bears the solar disk between his horns.

The practice was universal. Evidence of it has been found in the great civilizations of China and of India, in the tribal societies of Africa, throughout the Americas from Canada to Peru, and in Babylonia, too, where cauldron concerts join the lamentations of women: It seemed that everywhere the eclipses of the moon and of the sun caused pandemonium.

CHAPTER IV

COSMIC DISORDER

The appearance of the Donati comet in Paris in 1858 was later recorded in a lithograph (far left) by Amédée Guillemin. Left: This ancient Indian relief sculpture depicts Rahu, the demon of eclipses, holding a lunar crescent in each hand.

Although we know that solar and lunar eclipses are simply the passing of one celestial body into the shadow of the other, the sight of an eclipse still inspires wonder. And though even as early as the 1st century AD, the Roman scholar Pliny praised the astronomers who liberated human beings from the terror provoked by eclipses, he also recalled their fears: Some were convinced that a solar eclipse signaled the death of the sun; others believed the sun to be victim of an evil spell from which only "pandemonium" could release it. All over Europe, from Italy to Scandinavia, the custom persisted almost up until our own time. Each version is linked to the belief that at the time of an eclipse a monster or wild animal (a dragon, lion, wolf, or serpent) had attacked the sun or the moon and begun to devour it. The aim of the pandemonium thus seemed to be to frighten the devouring monster in order to force it to release its prey.

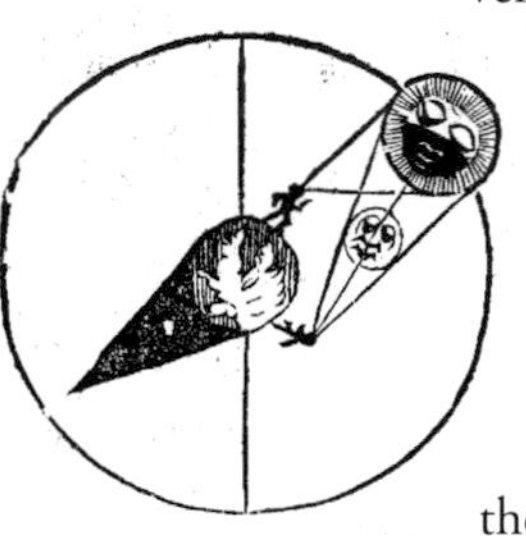

In the continual game of hide-and-seek that the sun and Earth play, Earth periodically comes between the sun and the moon in the course of its orbit. The shadow that it drags behind it then covers the moon and eclipses it (far right). The moon also sometimes travels between the sun and Earth, causing a solar eclipse (left).

Pandemoniums and Hullabaloos Served to Punish Threats to the Established Order

In their *Encyclopedia* French scholars Denis Diderot and Jean d'Alembert pointed out that similar events were often organized outside the homes of those who were marrying someone of an age greatly different from their own. Arnold van Gennep, in his monumental *Manual of French Folklore*, gives details of this custom and notes that generally it was the young people of the village who enforced the practice by organizing various types of noises called hullabaloos and by collecting fines whose amounts depended on the disproportion between the ages of the couple or the degree to which they had offended their fellow villagers.

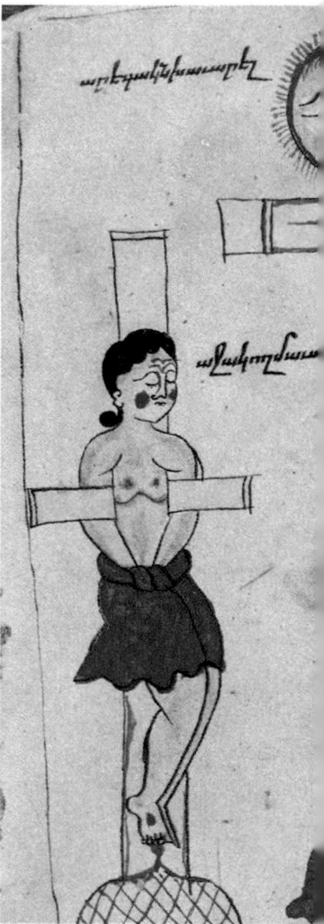

Scholars of folklore have imagined that hullabaloos were aimed at warding off unlucky influences acting upon a newly married couple and have compared the hullabaloos to the pandemoniums that accompanied eclipses, which were thought to consist of a dangerous union between a ravenous monster and a celestial body. The pandemonium and the hullabaloo drove away a "cosmological" monster in one case and a "sociological" monster in the other. In his analysis of the two events French ethnologist Claude Lévi-Strauss concludes that they were in fact aimed at the breaking of an order. In the case of eclipses, "It is the disruption of an order causing, by a regular chain of events, the alternation of the sun and the moon, day and night, light and darkness." In the case of marriage, the social order is

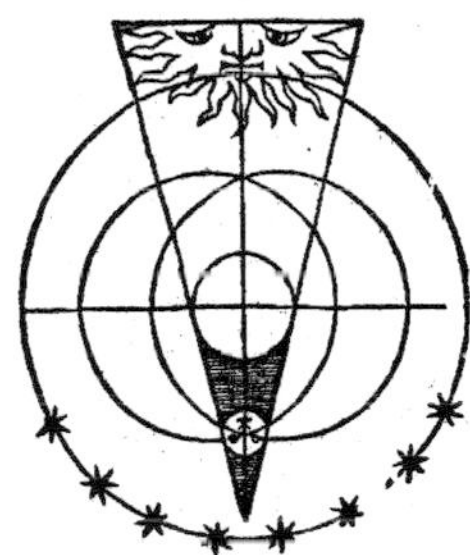

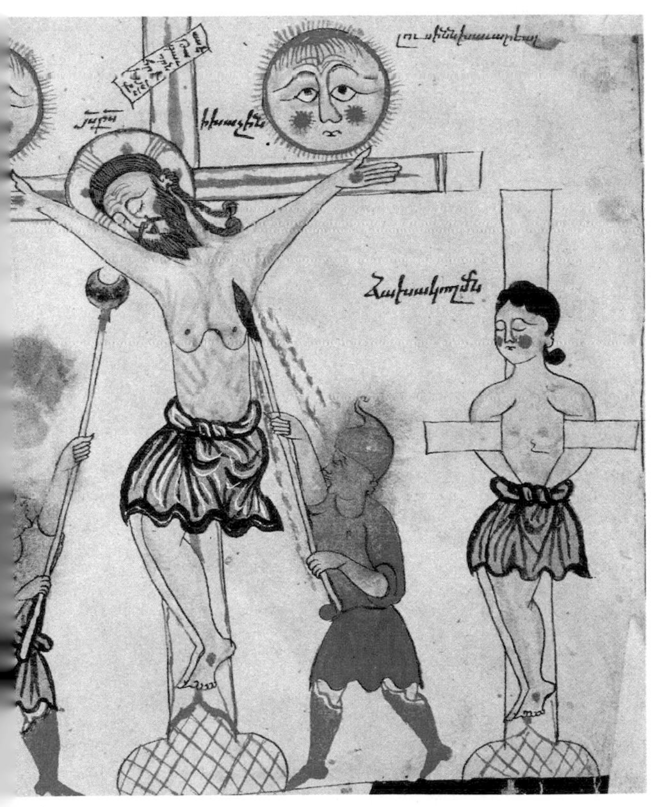

In this medieval Armenian miniature, the exceptional event of Christ's crucifixion coincides with an exceptional situation in the heavens: the simultaneous appearance of two suns in the sky. The Gospels of the New Testament reported an unexpected and abnormally long eclipse of the sun on that day.

The two luminaries shown in this 17th-century colored engraving reflect the Western practice of anthropomorphism, endowing nonhuman objects with human characteristics. Here the sun, who is portrayed as a man, embraces a moon radiant with feminine charm.

disrupted by disproportions "in terms of the social norms, whether related to age or wealth."

The fact that eclipses were often held responsible for the outbreak of epidemics seems to confirm Lévi-Strauss' interpretation. In South America it was believed that a solar or lunar eclipse heralded illness. When the sun was hidden it was said to be a sign of smallpox. And when in 1918 the Spanish influenza caused the death of so many South American natives, the people blamed a solar eclipse, "whose lethal venom had been spilled on the Earth."

Incest, which also frequently bears the responsibility for diseases, is sometimes associated with eclipses. A good illustration of this appears in the previously mentioned Eskimo myth of the creation of the moon and the sun (see pages 56–7).

There might be a correspondence between symbolic values associated with an eclipse and those of incest. Incest, which is found at the very core of all populating myths, is a universal taboo. The most extreme case of a "disproportionate" marriage, incest is the ultimate sign

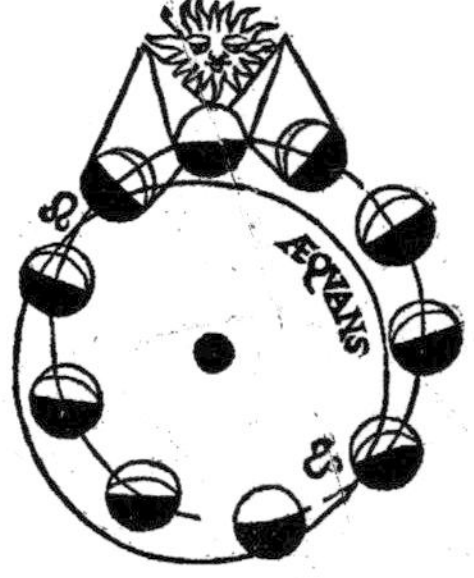

This illustration of the phases of the moon shows the points at which the orbits of Earth and the moon intersect. In Western mythology these points are referred to as the head and tail of the dragon. When the sun and moon occupy one of the points at the same time, the result is a solar eclipse.

During the 18th century Westerners were fascinated as much by exotic civilizations in faraway lands as they were by the mysterious happenings in the sky. Here is a contemporary engraver's idea of an eclipse in Peru and of the hullabaloo that went with it.

of social disorder just as the eclipse, the most extreme case of the phases of the moon or of the daily disappearance of the sun, is the sign of the greatest of cosmic disorders.

It's All the Comet's Fault

In his *Natural History* Pliny claimed that comets are stars that sow terror. The peoples of Ethiopia and of Egypt knew of a comet, to which Typhoon, the king at the time, gave his name. Fiery and spiraling in its movement, the comet appeared to be so terrible that it was perceived not so much as a star but as a ball of fire.

When the Inca king Atahualpa, who had been imprisoned by the conquistador Francisco Pizarro, learned that a great comet very similar to the one that had been seen shortly before his father's death had appeared in the sky, he was in the depths of despair. He was right to be fearful: The last Inca king of Peru was strangled on 9 August 1533.

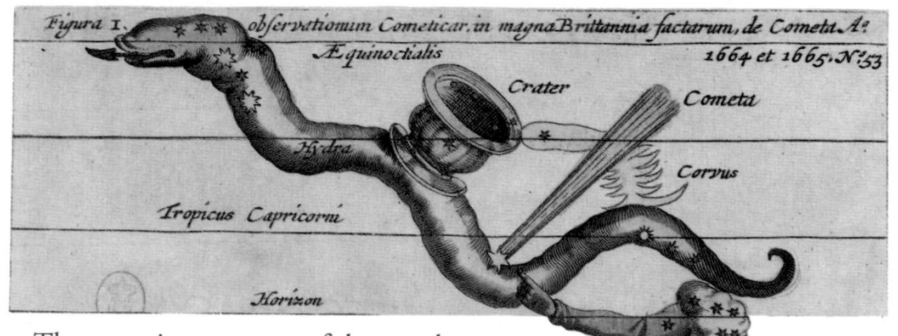

The accession to power of the murderous Roman emperor Nero was also marked by the appearance of a comet. His reign lived up to the terrible omen.

The Roman emperor Augustus was one of the few people who considered comets to be good omens. He went so far as to order that one of them be adored in a temple in Rome. This comet had appeared during the

In December 1664 a comet appeared in the constellation Hydra, the Sea Serpent. On its back are the constellations the Crater, or the Cup, and Corvus, the Crow.

games he had organized shortly after the death of his adoptive father, Julius Caesar. In Augustus' eyes the comet proclaimed that Caesar's soul had been welcomed into the company of the immortals, and he had a comet added to the bust of Caesar that was dedicated shortly afterward at the Forum. But Pliny claimed that the emperor rejoiced in the secret knowledge that the comet had in fact been born for him and it was he who was born in it. And it is true that Augustus' reign was (relatively) long and peaceful.

In the popular imagination a comet appears because the devil is lighting his pipe and is throwing away the still-burning match, which could be cause for worry. Therefore, one must watch the place at which a comet appears, the area toward which it is moving, the star that influences it, and the shape it assumes. If the comet looks like a flute, it is an omen related to the art of music; if it appears in the private parts of a

Beginning in the mid-16th century most of the ancient documents concerning comets that had been uncovered were assembled in "cometographies." The first of these was probably that of Swiss historian Johannes Stumpf, published in Zurich in 1548, but the most renowned and most beautiful cometography is Stanislaw Lubieniecki's *Theatrum Cometicum,* published in Amsterdam in 1667. One finds in it the history of all known comets from the time of Noah until 1665.

A shower of comets near Hamburg (left). Various apparitions and evil effects of comets (overleaf).

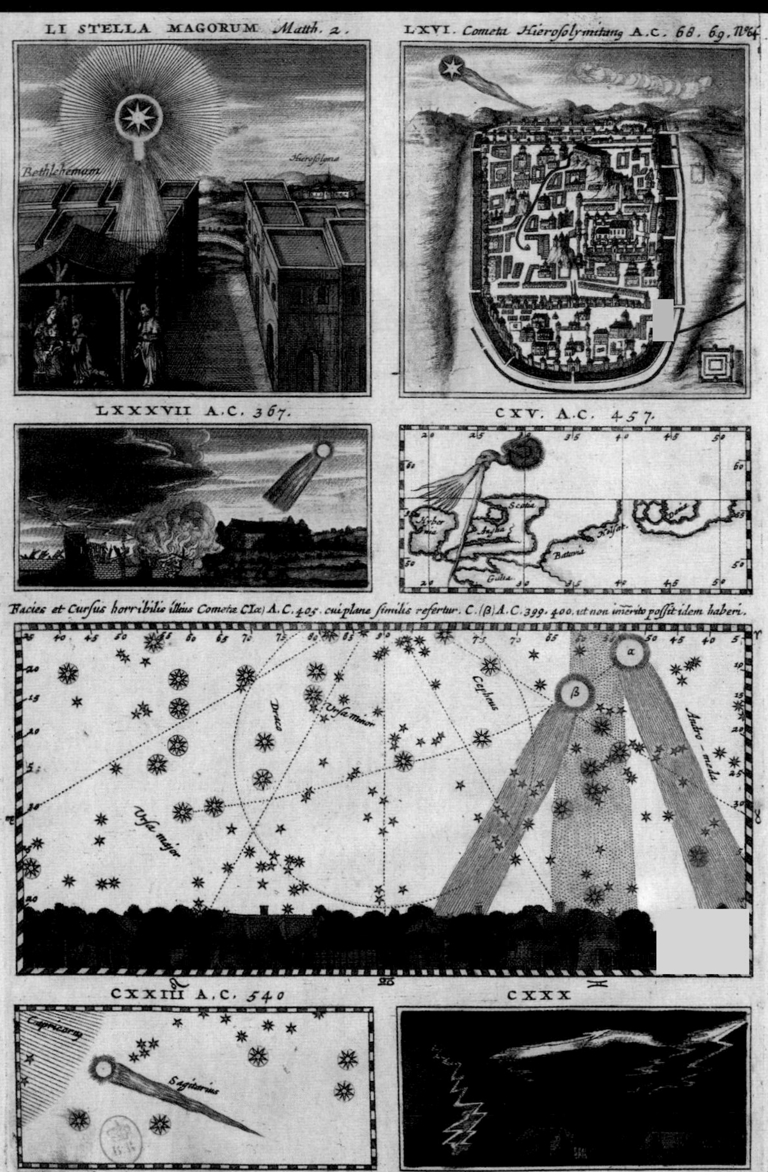
LI STELLA MAGORUM Matth. 2.
Bethlehemum
Hierosolyma
LXVI. Cometa Hierosolymitang A.C. 68. 69.
LXXXVII A.C. 367.
CXV. A.C. 457.
Scotia
Anglia
Batavia
Gallia
Facies et Cursus horribilis illeius Cometæ C(α) A.C. 405. cui plane similis refertur. C.(β) A.C. 399. 400. ut non inmerito possit idem haberi.
Draco
Ursa minor
Cepheus
Andro-meda
Ursa major
CXXIII A.C. 540
Capricorn
Sagittarius
CXXX
M. C. Ysenius del.
Stopendael sculp.

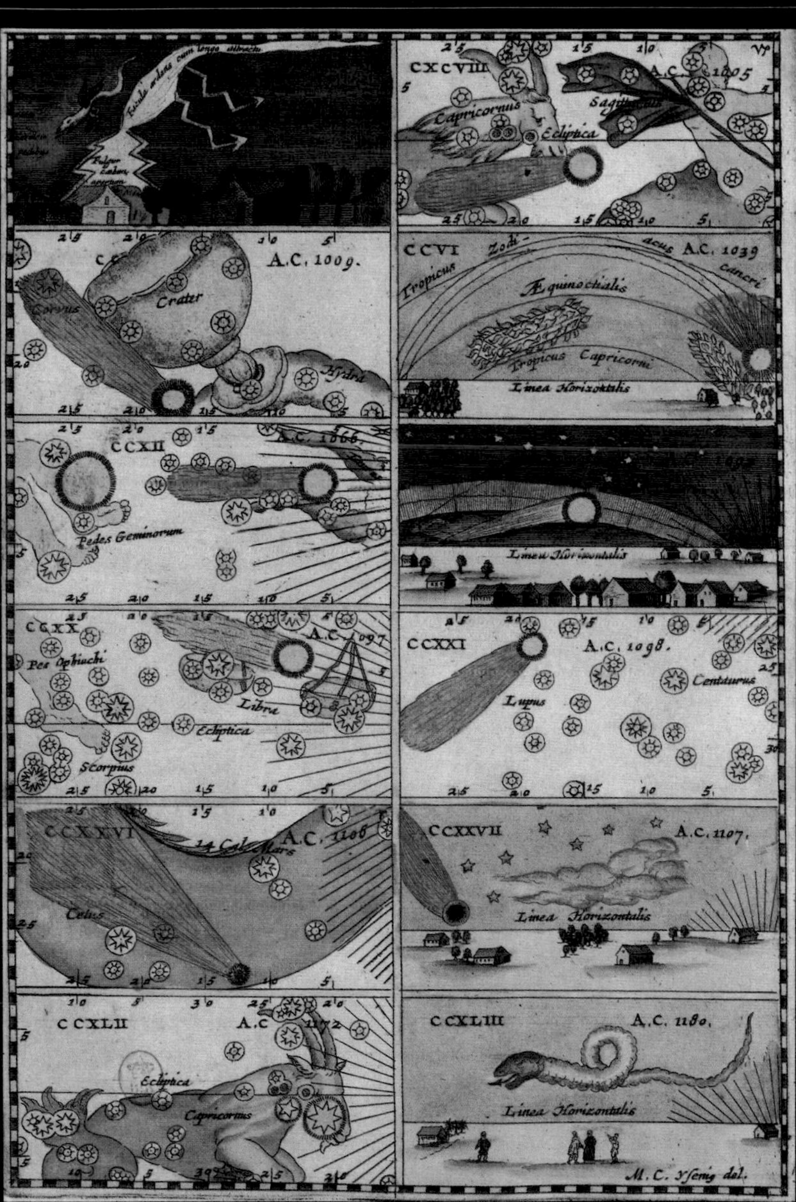
CXCVIII
A.C. 1005
Capricornus
Ecliptica
A.C. 1009.
Crater
Corvus
Hydra
CCVI
Zodiacus
A.C. 1039
Tropicus
Cancri
Æquinoctialis
Tropicus Capricorni
Linea Horizontalis
CCXII
A.C. 1066.
Pedes Geminorum
Linea Horizontalis
CCXX
A.C. 1097
Pes Ophiuchi
Libra
Ecliptica
Scorpius
CCXXI
A.C. 1098.
Lupus
Centaurus
CCXXVI
A.C. 1106
14 Cal. Mars
Cetus
CCXXVII
A.C. 1107.
Linea Horizontalis
CCXLII
A.C. 1172
Ecliptica
Capricornus
CCXLIII
A.C. 1180.
Linea Horizontalis
M.C. Ysenig del.

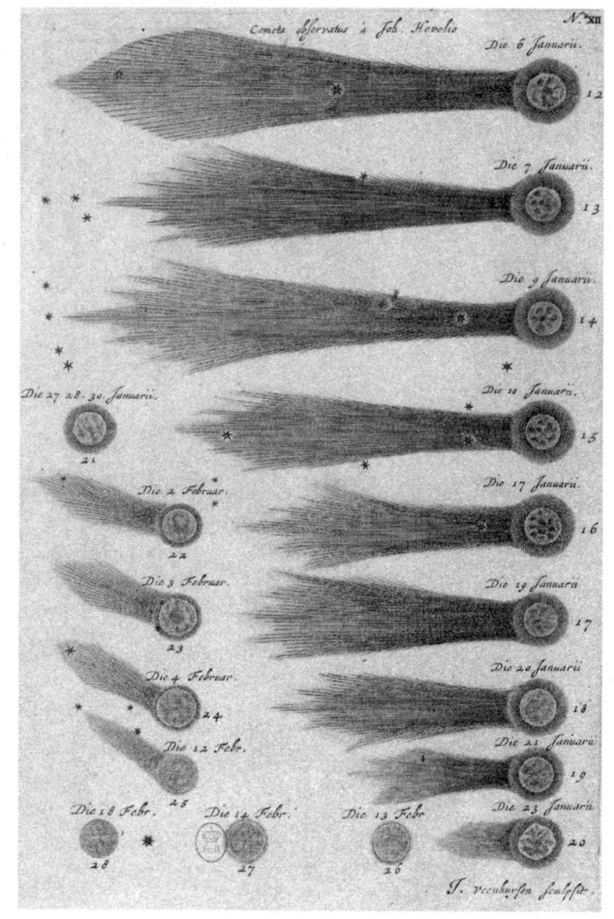

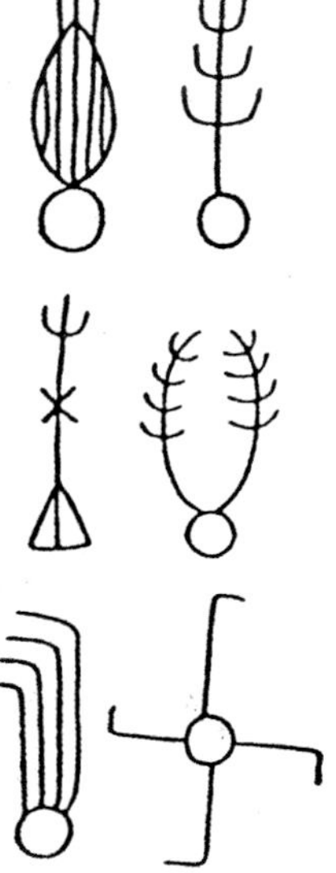

Comets come in many different shapes and sizes, and a wide range of them are included in the *Theatrum Cometicum* (above left and opposite).

constellation, then it concerns depraved manners, and if it forms an equilateral triangle with permanent stars, it heralds genius and wisdom.

For Many Years Scholars Were Divided and Uncertain about the Origin and Nature of Comets

The Greek philosopher Aristotle believed that comets were merely meteors born of the heating of Earth's atmosphere and spun between the moon and Earth,

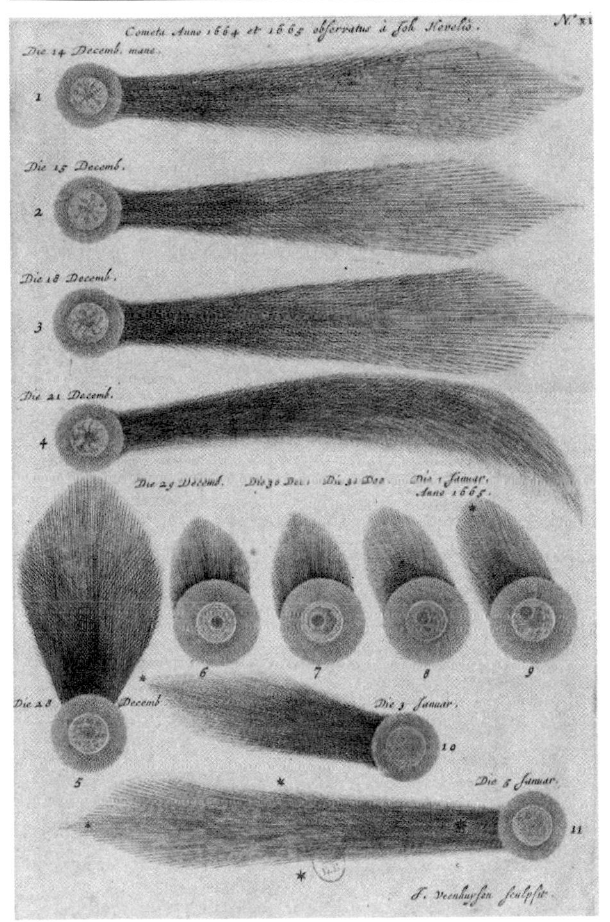

The schematized comets on the opposite page derive from the earliest known atlas of comets, the *Book of Silk.* Consisting of a silk ribbon about five feet long, it was discovered recently in China, in a tomb dating from the 2nd century BC. Its illustrations of twenty-nine comets, classified according to their appearance and the kind of catastrophe they herald, date from the 4th century BC. Apart from the forms of the comets, the book contains representations of clouds and of optical phenomena in the atmosphere.

whereas his 17th-century counterpart René Descartes considered them to be messengers from distant worlds. Today we know that a comet consists of a small, irregularly shaped nucleus of ice and dust, surrounded by an atmospheric halo, called a coma, and a tail of gas and dust. A passing star can alter a comet's orbit and bring it closer to the sun; as its temperature increases the comet begins to vaporize, and as it does its tail can expand behind it for hundreds of millions of miles.

Warhafftige beschreibung / was auff einen jeden sollichen Cometen geschehen sey / die gesehen
sind von anfang der Welt her / biß auff disen jetzgeschehenen Cometen in dem 56. Jar / auch waß sich an etlichen orten dar
nach verloffen hat / vnnd in welchem Jar ein jeder gesehen ist worden.
Es ist leider darzů kommen / das niemands weder auff wunderzeichen / noch auff geschichten etwas haltet / vnd jr niemands
war nimpt als ob sy vngeferd oder vmb sun also geschehen vnd gesehen werden. Nun finden wir in allen geschrifften das all-
Erinnerung vnd Warnung / von dem jetzt scheinenden Cometen
so in disem Monat Octobris / deß jetzt lauffenden 80. Jars / erstmals erschienen.
Mittag.
Auffgang.
Die erfarung gibts / das auff erscheinung der Cometen allzeit
natürlicher oder vnnatürlicher weise etwas erfolget. Dann
dern daß wirs am häytern Himel sehen sollen / darmit wir nicht mit
dem Gottlosen hauffen / das gespött darauß treyben / vnd dem Epicu

According to the 18th-century French naturalist Georges-Louis Buffon, a comet was responsible for setting the planets in orbit around the sun. Having searched through the archives of Europe, Buffon concluded that Earth and the other planets had been in a fluid state when they assumed their form and that, since this fluid could not have been produced by water, it must have been produced by fire. In contrast to certain of the comets, however, the fact that the planets had passed near the sun had not sufficed for them to become liquefied. He therefore assumed that this planetary matter had been projected out of the star in a single movement by the impact of a comet. He could point to no other body in the universe that could transmit such a tremendous movement to such great masses if it were not one of those comets that sometimes come so close to the sun that some of them must fall into it as they approach, streaking the surface and pushing ahead of it matter set in motion by the collision. Buffon also believed that comets were sources of the sun's energy.

These popular German prints from the 16th century (left) reveal the intense interest, even terror, that comets inspired. The passing comet in the upper engraving foretells an impending drama: The neighboring village will burn down.

The illustration below, from the *Theatrum Cometicum*, traces the path of the famous comet that first appeared in the sky at the end of 1664.

Comets Inhabit an Ordered World

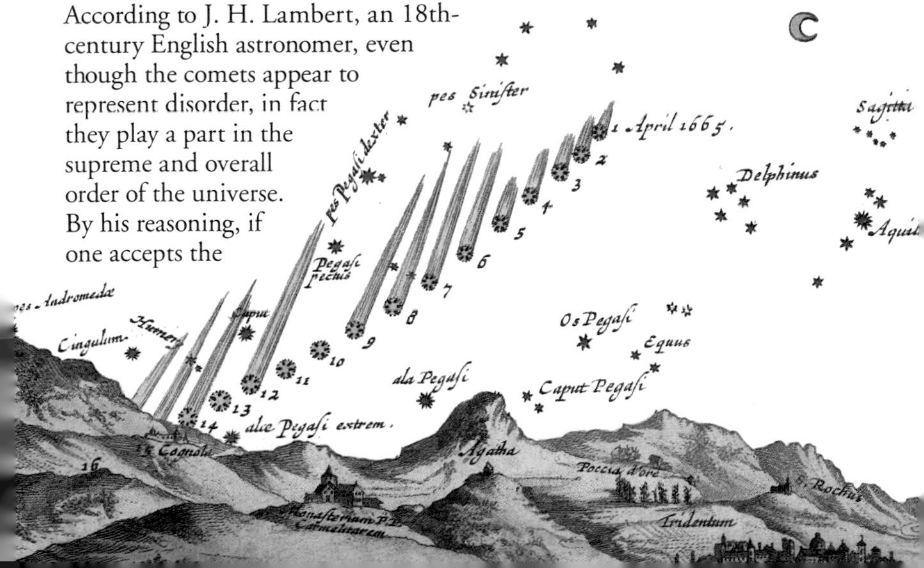

According to J. H. Lambert, an 18th-century English astronomer, even though the comets appear to represent disorder, in fact they play a part in the supreme and overall order of the universe. By his reasoning, if one accepts the

existence of a divine organizer who disentangled the chaos and gave the universe its form, then this universe is perfect and nothing happens in it by chance, not even the comets. Everything is done with intelligence, everything has its purpose, the means are subordinated to the ends, and some ends are subordinated to others. The principles of hierarchy, harmony, and plenitude reign in the world. If to some the universe has appeared disordered it is simply because they lacked a wider perspective and because their knowledge was imperfect.

As the scope of our vision has broadened and our knowledge increased, however, we have begun to see each of the heavenly objects at its appropriate place and distance, each one following a path from which it dare not stray and which seems to have been traced out for it with compass and ruler. It is then that the order and symmetry underlying the apparent disorder become apparent. There are within the universe innumerable heavenly bodies moving along set orbits, or paths, within its boundaries. The approximate makeup of our solar system is repeated countless times throughout the universe. Every star governs a world that is as full and as populated as our own and there exist as many of those worlds as the capacity of the whole universe that contains them can allow.

Contained in our small solar system are our star, the sun, and, revolving around it, nine planets, accompanied by their moons, which remain within the confines of the narrow band of the zodiac. From this brief description it might seem that space is sparsely populated. Yet added to this are thousands of comets with their varying orbits that fill up space. Also in the system are asteroids, made up of the rocky debris from the creation of the planets, which orbit together in a band around the sun. When asteroids collide, pieces of them, called meteoroids, can break off. When a meteoroid from space enters the atmosphere the heat produced by its passage causes it to flare brightly as it crosses the sky. This phenomenon is known as a meteor, popularly called a shooting star.

Meteor showers occur when the Earth, in the course of its orbit, passes through a swarm of meteoroids moving through space in a group. In mid-August the most famous shower of shooting stars, the Perseids, are visible. They are given that name because they seem to come from the constellation of Perseus. In fact it is Earth itself that at that moment is moving toward the constellation. The illusion, caused by relative motion, is heightened by the fact that Earth moves through space much faster than the meteoroids it comes upon.

The lithograph at left illustrates a shower of shooting stars seen in November 1872. Above: Sixteenth-century engraving.

A Silent and Fleeting Streak of Fire Crosses the Vast Space of the Night

Nowadays shooting stars have a beneficent reputation. When on a clear night a star seems to detach itself from the sky, move silently, and disappear, the custom is to make a wish. In earlier times, however, shooting stars were associated with souls, and their fall foretold someone's death or a change in the status of a deceased person in the afterlife. According to the *Gospel of the Distaff,* a best-selling chapbook containing hymns, poems, and legends, "One night, when you see a star fall, you must realize that it is one of your friends who has died, since we all have our own star in the sky and when we die, our star falls." In many regions a shooting star is a signal to say a prayer so that the gates of heaven may be opened to the soul of the deceased: The shooting star generally represents a soul that is going straight to heaven or whose period of penitence is over.

The Pilagas, a seminomadic tribe in northern Argentina, take a more prosaic view of shooting stars: They are the excrement of other stars. And some people say, more bawdily, that shooting stars are men who are in a hurry to get to that part of heaven in which women are waiting for them.

The appearance of a shooting star can be likened to that of a comet, but the phenomenon is more fleeting and less spectacular. An exception would be shooting stars that appear in showers, like the Perseids of mid-August. Each star in this group, which is sometimes referred to as the Tears of Saint Lawrence, the Judeo-Christian martyr who was burned on a grill, is said to represent a suffering soul who is appealing for the prayers of the living. Traditional greetings originally were supposed to be addressed to a suffering soul; today,

This huge ferrous meteorite was discovered in Sweden in 1870.

however, we seldom think about the deceased as we offer one another our own greetings.

Most shooting stars burn themselves out before they reach Earth's surface, a fanciful illustration of which appears in the lithograph above. Only the very largest ever reach the ground as meteorites.

When Objects Fall from the Sky

In July 1908 in central Siberia there came a sudden, deafening bang: An enormous boulder weighing forty thousand tons had just laid waste an area of the Siberian forest approximately thirty-six miles in diameter. If shooting stars are the grains of sand of the solar system, the Siberian meteorite was a gigantic rock. Fortunately, meteorites of that size are extremely rare. The oldest in recorded history is the one that fell in Thrace in 467 BC. Aristotle spoke briefly of meteors in his *Meteorological Treatise*, linking them with comets,

which, when dense and numerous, foretell a windy year. Shortly before the fall of the rock in Thrace a comet had appeared in the West. According to Aristotle the rock had not come from the sky but was simply returning to Earth, where it had been ripped from the ground by the wind earlier that day. Another Greek philosopher, Anaxagoras of Clazomenae, who is said to have foretold the arrival of this meteorite, believed that the stars were made up of white-hot rocks, a few of which occasionally became detached from the sky. It was in this disorder that Anaxagoras felt was revealed the true nature of the sky: It was rocky.

"[Jacob] dreamed that there was a ladder set up on the Earth, the top of it reached to heaven; and behold the angels of God were ascending and descending on it! And behold the Lord stood above it and said, 'I am the Lord, the God of Abraham your father and the God of Isaac; the land on which you lie I will give to you and to your descendants.'"
Genesis 28:12–3

Symbolic Stones Connect the Sky to the Earth

The large prehistoric standing stones called menhirs, and later the cosmic tree, the cross, and Jacob's Ladder are all symbols that form a link between earth and the sky. According to the Bible, on his way to Haran Jacob arrived by chance in a certain place and decided to spend the night there.. He picked up a stone from the ground, rested his head on it, and went to sleep. He then had the dream in which a ladder reaching from the ground up to heaven appeared to him. Afterward Jacob made a stela out of the stone he had used as a pillow, which had become for him a messenger from God. He called the sacred stone "bethel," which means "home of God."

Islam also has its "house of God," the Kaaba, a square building that contains the Black Stone, a famous meteorite. According to the Koran (2:44) Allah made the Kaaba, the Sacred House, the sacred month, and the sacrificial offerings with their ornaments, eternal

In the painting at far left, the boulder against which Jacob sleeps is mirrored in one of the clouds.

All Muslims are urged to make the pilgrimage to Mecca, where the Kaaba stone (illustrated in the 16th-century ceramic tile at left) is located, at least once in their lives: "Exhort all men to make the pilgrimage. They will come to you on foot and on the backs of swift camels from every distant quarter; they will come to avail themselves of many a benefit, and to pronounce on the appointed days the name of God."
The Koran 22:27

values for all humanity, "so that you may know Allah has knowledge of all that the heavens and the Earth contain, that Allah has knowledge of all things."

What makes the Black Stone so sacred is the ambivalence surrounding it: As a stone it is a terrestrial image; however, its celestial origins mean it is also a messenger from Allah. Through it Allah shows us that he knows everything about the sky and the earth. The Kaaba is the "center of the world." The place at which the Black Stone struck the ground as it fell from the sky is where the axis of the world passes, and the "center of the sky" is at its zenith.

It was said in Wallonia, in Belgium, that enormous lumps of rock roll around above the skies and set off storms. When two of these balls collide head on, lightning is produced by the impact, and the balls shatter into pieces. That is why on the days following a storm some searched for thunderstones lying in the fields. In fact it was most often because of their shape, which is in sympathy with the clouds or lightning, that the stones (which are supposed to have been brought by

The 19th-century engraving above reflects the popular belief that the stars were made up of stones, some of which could break free, either individually or in showers.

In this bronze statue from the 6th century BC, Zeus assumes the position of a javelin thrower to hurl the thunderbolts, of which he is master.

thunder or to have fallen with the rain) were chosen. Many arrowheads thus became thunderstones because of their resemblance to a bolt of lightning. In Sumatra, where a black cat presides over the rites destined to bring rain, a black stone whose shape recalls that of the cat is revered as the rain stone.

Thunder and Lightning Have Been Interpreted as Expressions of Divine Anger

Stormy weather has long been associated as a sign of the wrath of the gods. Ancient chroniclers report of the time that a torch shone suddenly in the sky, crossing the sky at noon before the eyes of the people during a contest of gladiators organized by Germanicus Caesar in ancient Rome. At Cnidus, in Asia Minor, in 394 BC a wooden beam fell from the sky, casting its sudden bright light during the naval defeat that cost the Lacedemonians the Greek empire. And in 349 BC, while King Philip of Macedonia was overrunning Greece, the skies suddenly opened with a great crack, raining down awful fires on the land.

According to Saint John (Revelation 6:12–4), at the end of the world the stars themselves will come crashing down through the opening: "When he opened the sixth seal, I looked, and behold there was a great earthquake; and the sun became black as sackcloth, the full moon became like blood, and the stars of the sky fell to the Earth as the fig tree sheds its winter fruit when shaken by a gale. The sky vanished like a scroll that is rolled up." At the end of the world the

Worn as an amulet, this polished and perforated pebble was supposed to protect its wearer from lightning. This thunderstone was probably shaped by the sea.

sky will once again open up and armies, angels, and fire will descend from heaven to destroy the Earth.

The Judeo-Christian apocalyptic vision mirrors the way in which the world of the Aztecs came to an end. The world had been placed under the protection of Tlaloc, the god of rain as well as fire; as Tlaloc fell from heaven he spread a destructive conflagration whose arrival was heralded by thunder and lightning.

Fortunately, however, the wrath of both gods and devils, of which thunder is a manifestation, is not always that extreme. Thunder can also serve as a calling to order, and in the past people have claimed the power to control it. Throughout Europe, and especially in southern Italy, priests and monks assumed that power until only recently. These organizers of storms had convinced the peasants of their ability to straddle clouds, positioning them over fields and causing them to rain on the crops below.

Ex-votos, paintings, inscriptions, or objects that are hung up when a prayer is answered, cover the walls of many country chapels. Here, the Virgin is being thanked for having spared the house, the cattle, and the harvest from lightning. This ex-voto also illustrates a Norman belief according to which whoever dares to look at the sky where lightning has torn it sees the figure of the Virgin in a corner of Paradise.

Saint Donat is reputed to provide protection from thunder, lightning, and hail. In the medieval chronicle of all the saints, the *Golden Legend*, Jacobus de Voragine reports that the prayers of Saint Donat, whom the pagans had held responsible for a drought lasting three years, caused the much-needed rain to fall.

The image at left is one of a series of popular prints from Epinal, which were accompanied by a prayer: "O merciful and eternal God, who holds all the elements in your mighty hands, and who governs the universe.... We ask you, restrain and stop all the infernal powers and push the terrifying rays of fire and of the tempest away from our heads."

When the Sky Catches Fire

From its highest reaches down to our own level, the sky can thus be filled with wonders. Related to these wonders are marvels, those rare phenomena that foretell an exceptional event. In extraordinary circumstances these miracles can cause further miracles to happen. We owe the story of the finest accumulation of marvels that history has left us to the Roman historian Livy, who lived from 57 BC to AD 17. While Hannibal, the great warrior from Carthage and the most formidable of Rome's adversaries, was preparing to leave his winter quarters and to resume combat, the anguish felt by the beleaguered Romans was intensified by a political crisis brewing in their capital city. Livy reported:

"These fears were revived further by the news that a number of marvels had occurred all around. In Sicily several soldiers had seen their spears go up in flames. The disk of the sun appeared to have shrunk. In Praeneste burning stones had fallen from the sky. In Arpi they had seen weapons floating in the air and the sun fighting with the moon. In Capena in the middle of the day, two moons had risen simultaneously. The waters of the Caere flowed with blood and even those of Hercules' spring were discolored by it.... In Capua the sky had caught fire and the moon had

The *loubok* belongs to Russian folk imagery. These broadsheets were sold at markets or door to door from the 17th to the 19th century. They were engraved and colored by self-taught artists who designed them specifically to appeal to the masses. The *loubok* below portrays an extraordinary storm, witnessed at Cartagena, Spain, during good weather, on 28 December 1743.

been seen falling with the rain. Less important marvels were also reported: The skins of some goats had turned to wool, a hen was turned into a cockerel, and a cockerel into a hen." Fortunately, a few pages earlier, just prior to listing a long series of marvels,

Another *loubok* illustrates an apparition in the sky, which was visible during an entire night in 1736.

Livy had warned: "Many marvels took place that winter in Rome and in its surroundings, or rather, as always happens when minds are inclined to be superstitious, many were reported and people hastily accepted them to be true."

When the biblical patriarch Enoch returned from his ascent to the uppermost levels of heaven, he confided to his children both secrets he had seen and those that the Lord himself had told him. Accompanied by an angel as far as the sixth heaven, Enoch had counted and measured the movements of all the stars. He also had counted the rays and witnessed the entrances and departures of the sun as well as its daily and monthly movements.

CHAPTER V

WONDERS OF THE SKY

When Flemish artist Pieter Brueghel painted the *Fall of the Rebel Angels* in 1562 (left), he wanted not only to show the disorder of the sky at the beginning of Creation, but also to attack the human follies of his time. Right: Aeolus, god of the winds.

Enoch had explored the area of the sky occupied by the clouds and their raindrops. He could describe the roar of thunder and the marvelous patterns of lightning. He had visited the stores of snow and the reservoirs of ice and cold air, and he had seen how the guards filled the clouds without ever emptying the warehouses. He had entered the hall of the winds and knew how their jailers supplied scales and measuring tools. He saw them first place the winds onto scales and then onto measures and then release them over all of Earth, carefully avoiding any sudden gust so strong that it could tip the world over.

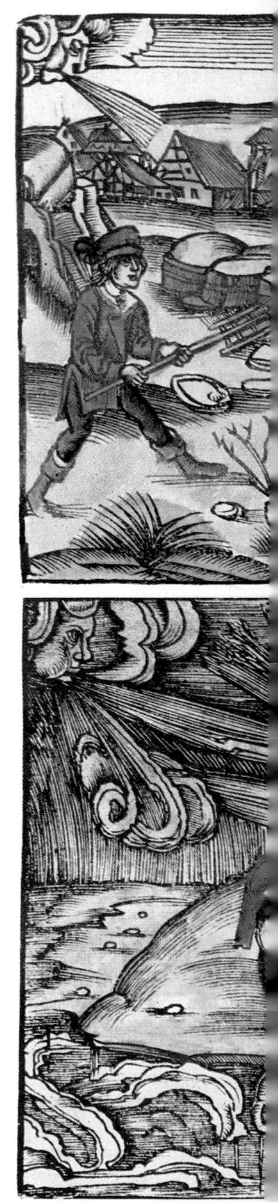

The Sky Is Not Only Where the Stars Shine or the Clouds Float. It Touches the Tops of the Trees and the Tufts of Grass at Our Feet

The sky is not restricted to the heights at which the immutable ether holds the stars under its circular law. It also includes all the areas of the atmosphere in which the breath of life from the heavens and the exhalations from Earth combine in infinite and ever-changing proportions. In the sky are born numerous evils for Earth and its inhabitants, and the threat of disorder is ever present. Nature struggles against herself there, and humanity is confronted directly by the celestial powers. Over our heads the jealous winds fight, fall to the ground, and return to the sky loaded with water, sand, and stones. The clouds rise and fall again as rain, snow, or hail while elsewhere the burning rays of the sun dry up rivers, lakes, and marshes. The ancients, who were convinced that Earth was immobile, believed that the daily rotation of the sky, which must have dragged with it the upper layers of the atmosphere, perpetually increased the disorder.

Earth's Order Is More Fragile and Easily Threatened Than That of the Heavens

Human beings have always gazed at the sky, which is so directly linked to their survival and their well-being. They have also populated it with a mass of divinities.

These miniatures illustrate the *Georgics*, Virgil's hymn to the glory of suffering humanity. This poetic treatise, written in the 1st century BC, is divided into four books: the corn and the ploughman's seasons, the vine and the olive tree, cattle rearing, and the beehive. The first book, illustrated here, deals with the cultivation of cereal crops. The ritual invocation to the divinities that protect agriculture is imbued with a religious fervor that suited the simple farmers. It was to them that Virgil directed his advice concerning the best way of cultivating their land to keep it fertile. He reminded them that heaven regulates the movements of the stars from above, influencing the sowing, growth, and harvesting of crops. This book, whose disjointedness likens it to an almanac, ends with a wonderful section on practical meteorology. The upper engraving illustrates the work of the ploughman; the lower one shows the natural calamities that afflict farmers.

Homer gave names to only four winds, which corresponded to the four points of the compass. The system was too general for the philosophers of the following age, who added eight more. This new system was judged too subtle and fragmented by Pliny who, convinced that the winds did not belong to the forces of disorder, kept only eight of the best known: From the East comes the *subsolanus*, at the time of the equinoxes, and the *vulturne*, at the time of the winter solstice; from the South comes the *auster;* from the West blows the *africus*, at the time of the winter solstice, the *favonius*, at the time of the equinoxes, and the *corus*, the Greek Zephyr, at the time of the summer solstice; from the North come the two snowy winds, the *septentrion* and the *aquilon*.

The illustration of the winds at left is taken from a 16th-century illustrated manuscript of Virgil's *Aeneid*.

Disruption of the regular course of the seasons upon which people depend has often been attributed to the gods' anger at human failings. Around 390 BC, after a terrible winter of snow and ice, the people of Rome were further tried by the rigors of an exhausting and fateful summer. In desperation the *duumvirs*, who were in charge of the sacred ceremonies, consulted the Sibylline books and advised that in order to appease Apollo, Mercury, Diana, and Hercules three beds should be set up for them for eight days. To accomplish this, reported Livy in his *History of Rome,* "All over the city, the doors of private homes were left open. All that one owned was placed out in the open and made available to everyone."

Regardless of differences among various systems of belief, all cultures incorporate the concept of the struggle between the forces of order and the forces of disorder as central to the very dynamics of life itself. Many projections have been made for the time when this struggle ends. According to the Gospels of the New Testament, the last day of time, Judgment Day, will be without sun, moon, or star; without clouds, thunder, or lightning; without wind, water, or air; without darkness, without evening or morning; without summer, without spring or heat; without winter, without frost or cold; without hail, without rain or dew; without noon, without night or dawn. It will be a day of death.

The Assyrian demon Pazuzu was the personification of the burning hot wind that comes from the south, bringing storms and fevers.

Aeolus and His Goatskin of the Winds

The wind, one of the most common signs of disorder, has inspired some very interesting legends, especially concerning its origins. In Brittany it is known that the winds first came to the sea, where at one time they did not blow as they do today. One legend tells of a sea captain who set off alone to explore the land of the winds. He brought the winds aboard his ship in well-sealed bags. Without telling his men what the bags contained, the captain gave strict orders that they should not be opened. He was being unrealistic. One night a sailor could not resist the temptation of opening one of the bags: Surouas, the southwest wind, escaped and blew so strongly that the ship broke apart, whereupon the other winds escaped from the bags, which had split open, and have blown over the sea ever since that time. This story is a distant echo of the one in which Aeolus, the Greek god of the winds, trapped them all in a goatskin, which he offered to Odysseus as a token of friendship. Curiously, Aeolus had gathered in the goatskin all the winds except the one that would take Odysseus home to Ithaca. Odysseus' sailors, like those of the Breton captain, eventually opened the goatskin, unleashing a tempest that pushed the ship back toward the coasts of Aeolis, where Aeolus lived. This time, however, a disgusted Aeolus washed his hands of Odysseus and his crew, leading them into further adventures, as recounted in Homer's epic poem the *Odyssey.*

This figure affixed to the face of a compass is called the rose of the winds (above left). It is divided into thirty-two sections representing areas of the winds and giving their cardinal and collateral points.

The Winds Are Vivid Characters That Inspire Insults As Well As Entreaties

The winds are very often personified, and in their almost-human guise they can have the same weaknesses as people. They have at times been called jealous, cowardly, or capricious. In his *Folklore of France* Paul Sébillot recounted a scene that he witnessed in 1880, which showed him that such attitudes continued to survive: On a ship from Newfoundland, which had been delayed considerably by strong wind, some men spat in the direction from which the wind came, shouted insults, and pointed their knives at it, threatening to cut it to pieces. Small children followed

Boreas was the Greek god of the north wind. He lived in Thrace, which the Greeks considered the land of cold weather. He was often shown as a winged demon of great physical strength. A member of the race of Titans, he abducted Orithyia, the daughter of Erechtheus, the king of Athens, and took her to Thrace, where eventually she bore him two children. Boreas was also the wind of fertility. With the mares of Erichthonius, he engendered twelve foals and later, with one of the winged creatures known as Harpies, many fast horses. In 17th-century alchemical iconography Boreas traditionally was portrayed with a newborn child in his abdomen.

the sailors' example, making the same gestures and repeating the same insults.

The women of Le Croisic in France preferred to mollify the winds rather than to insult them: On stormy days the sailors' wives would go to the chapel of Saint Goustan to beg the heavens to spare their husbands. After making their devotions they swept the floor of the sanctuary, collected the dust, and threw it into the air in the direction from which the wind had to blow to direct their husbands to the right port.

The Winds Continue to Have the Power of Titans

The fact that the winds are often personified as giants is of course due partly to their strength but above all to their role as forces of disorder. The ancient Greek poet Hesiod wrote of the race of giants known as the Titans, the sons of Heaven and Earth, whose rebellion against their father, Uranus, left the world in chaos. Things remained in disorder until the defeat of the Titans and the reorganization of the world by the Olympian gods, led by Zeus. And in the Bible, when Enoch describes his visit to Heaven, he places the home of Wisdom and the home of Violence alongside those of the stars and of the winds.

There are many giants in the myths of the Nordic peoples. To the east of the realm of the demons lies the Forest of Iron, where the giants who assume the forms of wolves live, and farther north lies the realm of the dead, where the giants of the frost and the eagle Swallow-His-Prey are to be found. It is the beating of the eagle's wings that unleashes the winds.

In art the wind has often been personified as a breath, which could be furious or celestial. Along with the symbolism of violence developed the imagery of the breath of the Holy Spirit. The Bible, like the Koran, makes divine messengers of the winds. The representation below is part of an 18th-century fresco by Italian painter Giovanni Battista Tiepolo.

Venerable Clouds Float Above Our Heads

Clouds play only a small part in celestial myths and legends. One finds at most a few tales and a large number of sayings in which clouds are given colorful names, usually related to their shape and color. For example, the large white cloud that is sometimes seen, hanging motionless above all the others over the French island of Sein, is known as Old John's Bunch of Flowers.

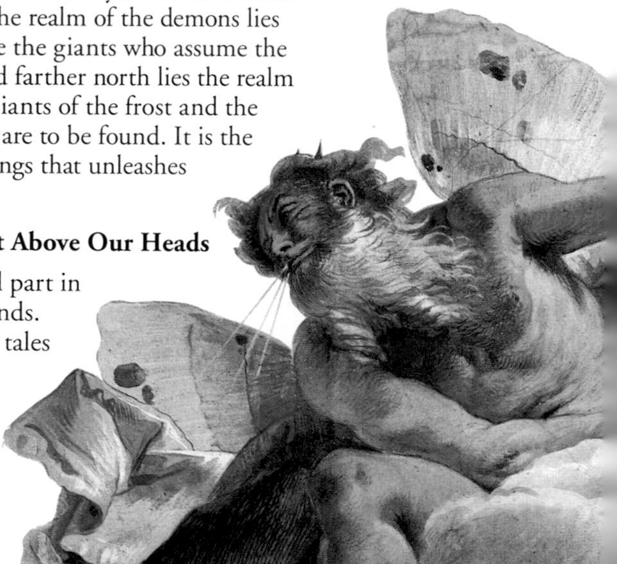

The large cumulonimbus clouds that one sees quite high up are often identified as trees, such as the tree of Saint Barnaby, the tree of Abraham, or the pear tree of the Maccabeans. While few clouds appear in the Bible, many trees are mentioned there, from the Tree of Knowledge to the Tree of Life to Abraham's Oak. In fact trees are often mediators between heaven and Earth. On the other hand it was from the top of a column of clouds that Jehovah spoke to Moses and his people. Moses had climbed the mountain of the Lord and the glory of the Lord remained there on Mount Sinai, hidden by a cloud, for six days. On the seventh day the Lord called out to Moses from inside the cloud. Moses went back down to his people and the column of clouds followed him; every evening it came to rest over the entrance to the tent that Moses had put up at the exit of the camp. At the sight of the motionless column of clouds at the entrance to the sacred tent, all the people got up and prostrated themselves before the tent.

Again in Enoch's apocryphal visions one finds the most intense evocation of a cloud: "There I saw the locked reservoirs from which the winds are distributed: a reservoir for the hail and the wind...for the fog and the mist, and the cloud coming from them has been floating over the Earth since the beginning of the world."

According to the Hindus, Vayu, which means wind, is the cosmic breath, the Word. It reigns over this intermediate world between the ether of the heavens and Earth. In ancient Persia the wind was the support and regulator of the world: "The first creation was a drop of water and it was from this water that all things were created.... Finally, the wind came in the shape of a fifteen-year-old youth who carried water, plants, cattle, just men, and all things."

The Celestial Ram, or Hook of the Clouds: The Rainmaker of the Dogons

For the Dogon people, the peasant-warriors who live a hard life in a chaotic setting amid the rocky landscape in western Africa, clouds provide the water necessary for their survival. It is understandable, therefore, that clouds are present in every aspect of their symbolic vision of the birth and evolution of the universe. In his book

God of Water Marcel Griaule describes a Dogon totemic sanctuary as a cube, each side of which measures about twelve feet. It has a facade flanked at each corner by towers that are conical in shape and crowned by navels, between which appears an iron hook, usually a double hook, each branch of which is bent in tightly. It represents the head of the celestial ram whose horns hold back the rain clouds. The two curled parts of the hook also refer to two hands that hold back the rain and gather in abundance. This celestial ram, which holds back the clouds in the spirals of his horns, is a golden ram. Before every storm he may be seen moving across the sky. The ram also represents the weather systems of the world. He urinates rains and fogs. He does not stay still as he does so, however; he runs through the clouds, leaving a trace of four colors of the soil, which he shakes from his hooves. That trace is the rainbow. The celestial ram uses the rainbow to come down from the heavens and dive into the great ponds on Earth. He dives among the water lilies and shouts, "The water belongs to me, the water belongs to me." The celestial ram is in fact the primordial cloud, whose first rain fertilized the first field, urinating the rains of the sky onto Earth. Thus, just as the wind carries the clouds, the clouds carry the rain, snow, and hail.

This gold pendant, with its curved horns to hold back the clouds, represents the celestial ram of the Dogons. It also symbolizes the anvil of the Blacksmith Ancestor, who forged metal in the first field.

The Seven Bright Stars of the Pleiades Are Almost Universally Associated with the Rainfall Pattern of Tropical Regions

Tropical regions, as well as those bordering on the equator, are subject to singular rainfall patterns, either where more or less heavy rain is always present or where

In Greek mythology the Pleiades were seven sisters, the daughters of the giant Atlas and Pleione, the daughter of the Ocean. They were called Taygete, Electra, Alcyone, Sterope, Celaeno, Maia, and Merope. All of the Pleiades married gods, except for Merope, who married the mortal Sisyphus and was ashamed of this. That is why the star that is dedicated to her is the faintest of the seven. It was because of Orion that the sisters were transformed into stars. The hunter fell in love with all seven sisters and pursued them for five years. Zeus took pity on them and changed them first into doves and then into stars.

rainy and dry seasons alternate regularly. The distinctive cluster known as the Pleiades, which is a part of the constellation Taurus, has inspired a host of related legends.

The Indians of French Guiana tell of how seven gluttonous sons, whose mother refused to feed them continuously, decided to change into stars. As the Pleiades they now reign over the rain.

In some areas the disappearance of the Pleiades behind the western horizon marks the end of the rainy season and the period of the greatest festival of the year. In other areas their disappearance in May and reappearance in June herald floods, the molting season of birds, and the renewal of vegetation. In French Guiana the Pleiades' return to the horizon is welcomed enthusiastically by the natives because it heralds the beginning of the dry season, whereas their disappearance is accompanied by a return of the rains that make travel by boat impossible.

The fact that the chronological alternation of the appearance and disappearance of the Pleiades coincides with the alternation of the rainy and dry seasons is sufficient to single out that star cluster as responsible for the rain. The tropical peoples, therefore, have interpreted the occurrence as a causal link as opposed to a coincidence.

Water traditionally is seen as the source of life, a means of purification and a place of regeneration. Fairies often were the spirits of feminine waters, and they liked to inhabit the area around a well, such as the one illustrated in this 14th-century woodcut.

The Symbols Associated with Water Are Numerous, Complex, and Contradictory

Like fire, water possesses dual characteristics and has a dual significance. Waters come both from above and from below. There are waters of life and waters of death. While beneficent fairies are thought to take refuge in the living water of springs, the stagnant water of ponds are said to harbor the devil.

Water can truly be the water of life. The writings of Herodotus, a Greek historian of the 5th century BC, tell of emissaries asking the Ethiopian king about the lifespan of his subjects; the king replied that in his realm people lived to the age of 120 years and longer. When the envoys looked surprised he led them to a spring whose waters gave off a scent of violets and made one's skin smooth. This water was so light that nothing, neither wood nor materials lighter than wood, could float in it; everything just sank to the bottom. The Ethiopians owed their amazing longevity to the fact that they drank that water.

In contrast to the water of life that healed, rejuvenated, or even guaranteed eternal life, there also existed a water of the dead. "To become water signifies death to the soul," said the philosopher Heracleitus, who lived about a century before Herodotus. In ancient Greece it was believed that the dead were thirsty in the period before the spring rains. During certain ceremonies water was poured out to the dead through crevices.

In his *Germania* of the late 1st century AD the Roman historian Tacitus spoke of this other aspect of the element as he described how the peoples of the North honored the earth-mother:

"There is an island in the ocean, a holy forest, and there there is a consecrated chariot, covered with a veil: Only the priest is allowed to touch it. He knows that the goddess is present in her sanctuary and he accompanies her very respectfully, pulled along by heifers.... Then the chariot, the veils, and, if one is to believe the narrator, the divinity herself, are bathed in a secluded lake. Slaves perform the task of washing her and as soon as they have finished, the lake swallows them up."

There are many documents describing the sacred peat bogs of Denmark, Norway, and Sweden. In Iceland there were many *blotkeldur,* or sacrificial marshes, into which men who had been hanged were thrown as sacrifices to the gods. The Tartars threw their illegitimate children into the mud on the banks of their sacred ponds. Until quite recently in Cornwall, England, sick children were immersed three times in the well of Saint Mandron to emerge from it cured.

The Dogons know that not every source of water is pure. The seventh mythical ancestor of their culture had vomited precious stones and water that contained impurities. This polluted water was spread over the Earth in ponds and rivers, and it was at that point that the ram urinated the first rains, which purified the water.

"My lover has the virtues of water: a clear smile, flowing movements, a pure voice that sings, drop by drop."
Victor Segalen

Tlaloc, "he who makes things grow," was the Aztec god of rain. He was assisted by minor gods who distributed the rains. Tlaloc lived in a garden, the symbol of abundant vegetation.

The Symbolic Ambiguity of Water Is Reflected in the Virtues and Powers Attributed to Rain

Rain, like spring water, has many curative properties. "Water truly heals; water expels and cures all diseases," asserts the *Atharva Veda,* one of the great religious texts of India.

At the outbreak of a storm, folk tradition in western Brittany calls for rheumatics to strip and lie face down, exposing their backs to the downpour until the storm's end. Drops of rain that fall on Saint Lawrence's Day are a certain remedy for burns.

Like underground water, rainwater has fertilizing properties. A Melanesian myth tells of how a young girl lost her virginity because she allowed the rain to flood her body. A related myth relates that another girl became a woman when she was touched by a few drops of water that had fallen from a stalactite.

Before planning a wedding in France one should check both the weather forecast and local folk traditions. At Dinan in western France rain falling on a wedding day tells the bride that she will be happy because the tears she would have otherwise shed have

Because of the composition of its riverbed, which absorbs water to the point of saturation and then releases it, the floods of the Seine are slow and long lasting. Before its course was regulated the river used to cause catastrophic flooding; the flood of 1910 has remained in popular memory. The anonymous oil painting shown here was copied from an illustration in the *Petit Journal.*

fallen on that day, whereas in the Poitou region rain is a sign that the bride will be beaten and that she will shed as many tears as there are drops of rainwater. A rainy wedding also warns the poor Poitou bride that she will be the first of the pair to die; if the sun shines it is the husband who will be buried before her. Rainfall on a wedding day in Marseilles guarantees prosperity in the household of the couple, but in the Vivarais region it heralds poverty.

The colored woodcut below illustrates Jehovah's instruction to Noah: "Go into the ark, you and all your household, for I have seen that you are righteous before me in this generation. Take with you seven pairs of all clean animals, the male and his mate; and a pair of the animals that are not clean, the male and his mate; and seven pairs of the birds of the air also, male and female, to keep their kind alive on the face of all the Earth."

Genesis 7:1–3

The Great Flood Is a Universal Symbol of Divine Punishment

In the majority of the myths from the Pacific rim it is a ritual fault, whose responsibility is shared by the entire tribe, that unleashes deadly floodwaters. Among the seminomadic farmers of the high Vietnamese plains, who "devour the forest" to provide fertile fields for their crops, one case of incest is enough to set off torrential rains.

Noah's flood was unleashed as a result of a collective sin, involving the whole of the human race. The Lord saw that men had been truly evil. The springs and the heavens were opened! Underground water and water from the skies allied their destructive and purifying powers to flood the Earth and return the world to its original formlessness so that Noah and the chosen survivors could begin anew.

As the waters rose everything that breathed died, except for the inhabitants of Noah's ark, who

Votive offerings, or ex-votos, are meant to express gratitude for a supernatural intervention. Their diversity is enormous, ranging from souvenirs linked to the miracle or manufactured objects purchased at the gates of sanctuaries to graffiti, paintings, drawings, or mosaics recounting the miracle and its circumstances. All are "signed" by the grateful individual. The most prolific category, votive paintings, employ various supports: wood, canvas, cardboard, paper. The oldest of them were painted in tempera on roughly planed boards. The subjects of these pictures encompass an exhaustive repertory of the human tragedy, including accidents, diseases, and natural disasters. Heavy rains have left a particularly lasting impression on the collective memory of traditional society, as is shown by this votive painting that describes the material damage and distress caused by floods.

The large number of ex-votos offered to chapels by those who survived the perils of lightning, storm, or flood could be explained partially by the terror that the abrupt break these events cause with the everyday order. A more profound interpretation is that accidental, sudden, or unforeseen death was so feared and the saints were so revered for protecting people from it because, in a rural society still very much attached to religion (even if the worst superstitions also played a role in it), the idea of dying without the last sacraments was dreaded as a true catastrophe. Here, some sailors have taken refuge from the storm in the cabin of Provençal customs officers. Lightning strikes the roof of the shelter and burns it down.

In Provence and Brittany, both maritime regions of France, one finds many ex-votos devoted to the dramas of the sea. The ex-voto often portrays the circumstances of the drama. Small coastal fishing boats were particularly vulnerable, but boats that went far from shore to fish, as well as large ships, were also at risk. Generally, the Virgin Mary is credited by those who have survived a storm. Some sailors painted their own pictures, using tarred sail canvas and painted from the ship. Marine ex-votos tell the stormy adventures of sailors and the sea, recounting the many dangers faced by the maritime travelers.

had found favor with God. They endured forty days and forty nights of rain. As the level of the water began to fall Noah released the dove; when the dove returned it held in its beak a fresh branch of an olive tree, a symbol of peace.

Thus, even as water destroys it also brings about purification and regeneration. In Genesis 8:21–2, as Noah and his sons left the ark, Jehovah said: "I will never again curse the ground because of man.... While the Earth remains, seedtime and harvest, cold and heat, summer and winter, day and night, shall not cease."

From this great disorder a new order is born, symbolized by the rainbow, which appears only after it has rained and which God unfurled over Noah's ark as a sign of a new covenant with the people. In many areas of the world the rainbow, which usually represents positive values, is referred to as the "Ark of God," or of a popular saint, most often of Saint Martin or Saint Michael.

Both in the West and the East the rainbow occupied an important place in the elaborately illustrated medieval "books of wonders."

The Rainbow Can Sometimes Display a Malevolent Aspect

The rainbow has also been called the Devil's Bow or the Wolf's Tail. In Celtic legend those who had seen a rainbow up close claimed that it had an enormous snake's head and flaming eyes. It is said that when the devil came down to earth in this menacing and terrifying guise, he dried up whole lakes in an effort to satisfy his limitless thirst.

One finds reference to this dark side of the rainbow as early as the time of Homer. Whereas the biblical rainbow symbolized the new order and covenant, Homer's rainbow heralded fresh disorder and catastrophe. In the *Iliad*, Homer's epic poem recounting the Trojan War, prior to the death of Achilles' comrade Patroclus, Zeus sent Athena to renew the quarrel between the Greeks and the Trojans and then spread the rainbow over the clouds for all people to see, which was his way of announcing the outbreak of war or a hurricane. Later, Zeus offered victory to the Trojans

The two lines of Arabic text framing the rainbow in this 13th-century miniature (left) describe the conditions under which a rainbow is formed, its colors changing with atmospheric conditions.

In this 18th-century engraving the Roman god Jupiter is surmounted by the two signs of the zodiac that he governs: Sagittarius, the sign of justice, and Pisces, the sign of philanthropy. From his chariot, drawn by two eagles, he hurls thunderbolts, and because of this practice he is invoked with the epithet Elicius, from the Latin verb *elicere* (to attract). It is he who attracts the thunder from heaven and enables it to come down to Earth.

while terrorizing the citizens of Argos by hurling bolts of thunder and lightning at them.

The Rainbow as Monstrous Serpent

The ambiguous symbolism of the rainbow is also found among the native tribes of South America, with one difference: Its positive value does not foretell a covenant but a break. As the herald of the end of the rain (in Genesis it also heralded the end of the flood), it marks the separation of earth and sky, which the rain had linked. The rainbow is born of rain, and both its ends rest in the mouths of the two serpents that cause rain to fall. Its appearance is a sign that rain has stopped, and when it disappears it is because the two serpents have gone up to heaven to hide in a pond there. They will come back into earthly water at the time of the next heavy rain.

In Australia, also, the rainbow is linked to the serpent and is held responsible for certain diseases. When the Europeans brought smallpox to the continent the aborigines called that disease "the scales of the Great Serpent." As one of the totemic ancestors, the Australian Rainbow Serpent is doubly symbolic: It is good and evil and evokes creation and destruction. It took part in the birth of the world; in some versions it only created the great rivers, and in others it is assimilated with the creative Great Mother. Its powers arouse so much terror that believers are advised to watch out for it. Pregnant women are advised not to pollute the water holes where the serpent goes to drink, and at the time of the rites that mark their entry into adulthood, young boys are cautioned to avoid drinking from the rivers, in case the Rainbow Serpent should abduct them.

The Rainbow Serpent plays an important role in the aboriginal culture of Australia. As one of the totemic ancestors, it plays a central part in the regeneration of nature and in human fertility. It is the force of both good and evil; the rainmakers and the medicine men can influence its malevolent and benevolent powers by manipulating objects from which those powers emanate, such as quartz crystals and shells.

The Myth of the Rainbow as a Path Linking Heaven and Earth Is Prominent in Australian Legend

The Australian gods reign in heaven seated on crystal thrones, and mythical heroes join them by climbing up the rainbow. The ascent to heaven by means of the rainbow is a very important phase in the initiation of witch doctors, a ceremony accompanied by symbolic deaths and resurrections. The master, who takes on the appearance of a skeleton, introduces the candidate, who has been reduced to the size of a small child, into the bag the master wears around his neck. Then, straddling

These representations of the Rainbow Serpent, painted on sheets of bark, are very common in the Arnhem, the region of northern Australia explored in 1623 by the Dutch navigator of the same name. There, under the name of Yulungurr, the Serpent lives in a sacred well called Mirrimina. It is the main character in the mythical history of the Waliwag sisters, who were linked to fertility. The sisters gave names to the animals and plants as they traveled through the country. When she was about to give birth to a child, the youngest of the sisters stopped near the lagoon in which Yulungurr lived. While building a cradle for the newborn baby the eldest sister inadvertently polluted the lagoon. The Rainbow Serpent was offended and unleashed a great storm to frighten the sisters, who, in an effort to calm it, started dancing and chanting the names of the animals. This only increased Yulungurr's anger and he smothered them in a violent storm.

the rainbow and using his hands as one does to climb up a rope, the master leads the candidate to the top of the rainbow and throws him into heaven. After letting some small freshwater snakes and quartz crystals into the candidate's body, the master brings the candidate back to Earth, again by way of the rainbow.

Throughout Europe, the Rainbow Is Associated with a Wealth of Images and Powers

Many sailors believed that if their ship happened to pass by one of the ends of the rainbow at the moment it was taking in water they would be swallowed in it. The notion that by passing under the rainbow one risked changing one's sex was commonly found in Renaissance comedies and stories.

Elsewhere it required more effort to change one's sex, especially for girls. In order to become boys, they had to throw their hats over a rainbow. In many places it was considered dangerous to point at a rainbow: At worst one could lose one's finger; at best it would become severely inflamed. The rainbow as provider of wealth—gold, silver, or, more often, pearls—is a commonly held myth; to gather a rainbow's treasure one must place a basket at its end, under one of the pillars on which it rests.

In his *Book of Wonders* published around 1557 Conrad Lycosthenes gave a precise inventory of the astronomical and meteorological phenomena that occurred between 2307 BC and AD 1556. Illustrated here is a paraselene, or the splitting of the moon in two, observed in 1168.

The Magic of Celestial Light Shows

The play of sunlight and, sometimes, of moonlight with the atmosphere, which is full of drops of water or small ice crystals, creates beautiful optical marvels, such as rainbows, halos, and parhelia, more commonly called sun-dogs.

The native Australians relate the moon's halo to the time that Balou, as the moon was known, came down to Earth. Unhappy with the avarice of the ibis

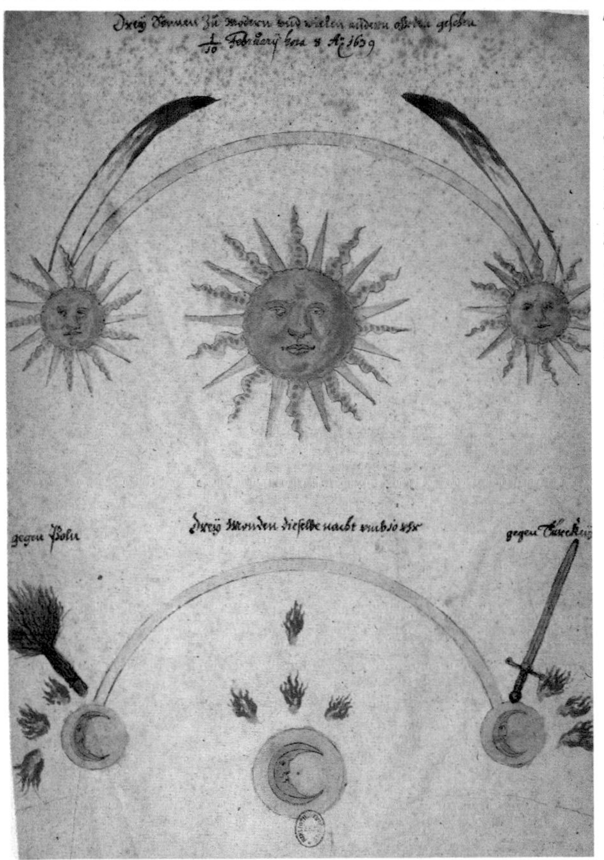

The splitting in two of the sun, a parhelion, and that of the moon, a paraselene, as Conrad Lycosthenes called them, are optical phenomena caused by the refraction of light in the small ice crystals floating in the upper atmosphere. This illustrator of the 17th century saw torches, flying sparks, and swords in the arcs that accompanied the divisions.

Mouregou, who would not welcome her in his house, and to protect herself from the cold, she built a small, circular hut made entirely of bright bark. During the night it rained heavily and Mouregou's house was flooded. Since then, whenever Balou the Moon appears in her round hut in the sky Australians have come to expect that it will rain the next day.

The sun may also be surrounded by a halo or a corona. According to Roman legend, when Augustus entered Rome after the death of his father, Julius Caesar, the sun appeared surrounded by ears of wheat and circles of colors. If the sun disappears during an eclipse, three suns will appear to rise afterward—the usual one accompanied by two sun-dogs. Exceptional cases have been recorded when the dogs of the sun followed it from early morning till evening. Explorers returning from the poles reported instances when they saw what seemed to be six suns rising simultaneously.

The Auroras Are Among the Most Magical of the Celestial Light Shows

Auroras have been described as a time when the night sky grows purple, when torches and lamps appear to float in the dark and nocturnal suns light up the night with a semblance of daylight. In reality auroras consist of streams of light in the upper atmosphere of Earth's polar regions caused when particles, or electrons, from the sun leak through the double layer of charged subatomic particles that surrounds Earth. Auroras can last for a few minutes or for an entire night. The peasants of the Beauce region of France and of Corsica in Italy

The aurora borealis, also called the northern lights, are formed by charged solar particles trapped in the upper atmosphere by Earth's gravitational field. This field acts like a magnetized bar, the ends of which are at the poles. The corresponding phenomenon in the southern hemisphere is referred to as aurora australis, or the southern lights.

reported an aurora borealis visible sometime before the Franco-Prussian War of 1870.

Toward the Last Land of Myths?

Since the dawn of time, all over the planet people have attempted to decipher the sky. But the sky, especially the distant sky, has long been discreet and secretive, and ancient peoples projected more images onto the sky than they received messages from it. From the tops of the trees to the home of the gods, they loaded the sky with a mass of symbols, myths, and legends, making it an integral aspect of the history of the world. To transform Callisto into a bear and project her into the heavens was to show her to everyone for all eternity, regardless of whether it was as a punishment or a glorification.

Peoples of the past believed that they could read the signs of the heavens and interpret them well before they had any understanding of the true workings of the sky. Since the time of Galileo, however, and with the increasing sophistication of astronomical knowledge, popular lore and beliefs surrounding the sky have steadily declined in the face of science. Now that the advances made by science are so spectacular and the broadcasting of discoveries is instantaneous and almost universal, and now that textbooks and schools are replacing the oral wisdom of village elders, the great myths and their images have lost their validity as fact, but the grandeur of the imagination they represent will remain with us always.

The moon does sometimes pass in front of Mars, the red planet, which is personified as the Roman god of fire and war. This celestial phenomenon has been fancifully interpreted in the illustration above. Overleaf: The southern sky, lithograph by French artist Amédée Guillemin, 1877.

Expose these maps to the light for a few minutes before going outside

Because of Earth's rotation around the sun, in every place on the globe the appearance of the sky changes slightly every night. The following maps show a few of the stars and constellations visible in the sky in March, June, September and December.

These maps are accurate for a latitude of about 45°. They approximate the starry sky as you will see it by turn[illegible]g either to the north or to the sout[illegible]n the late evening, at about 9 PM. Add one hour for daylight saving time.

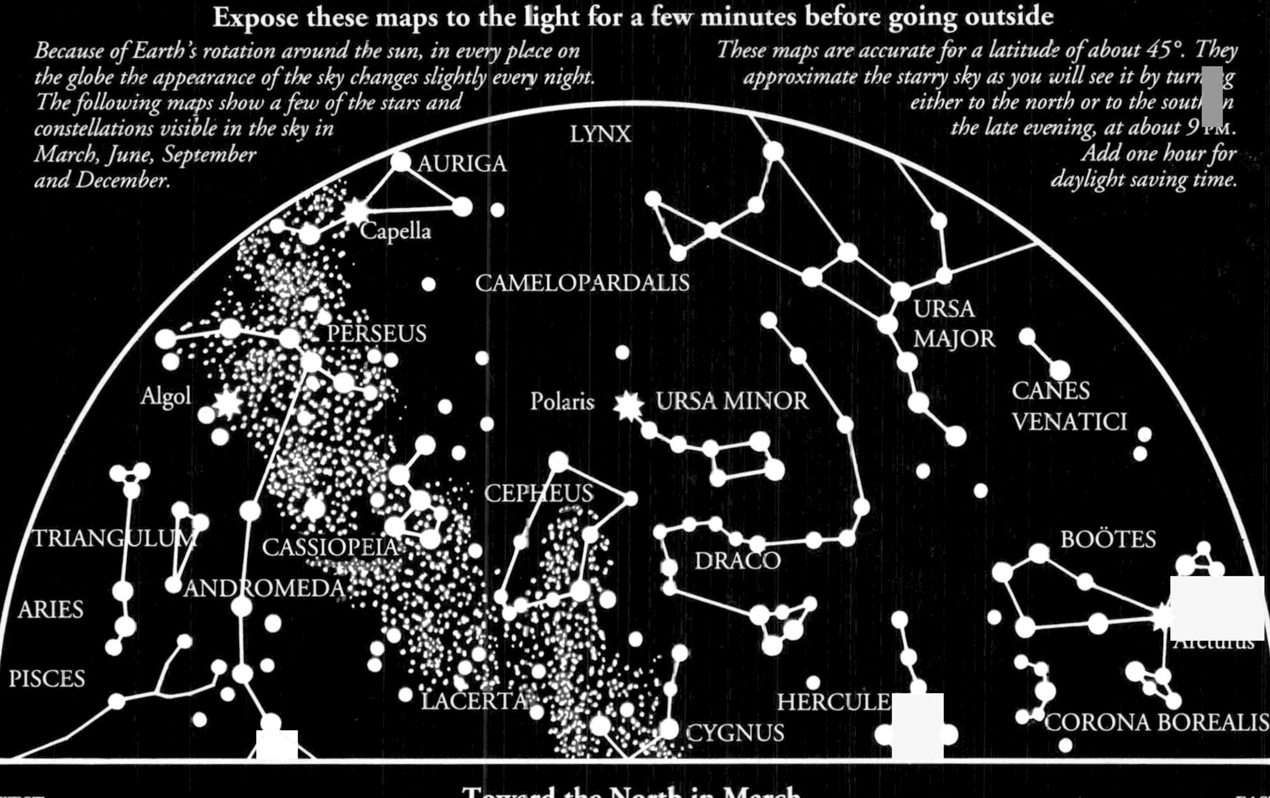

Toward the North in March

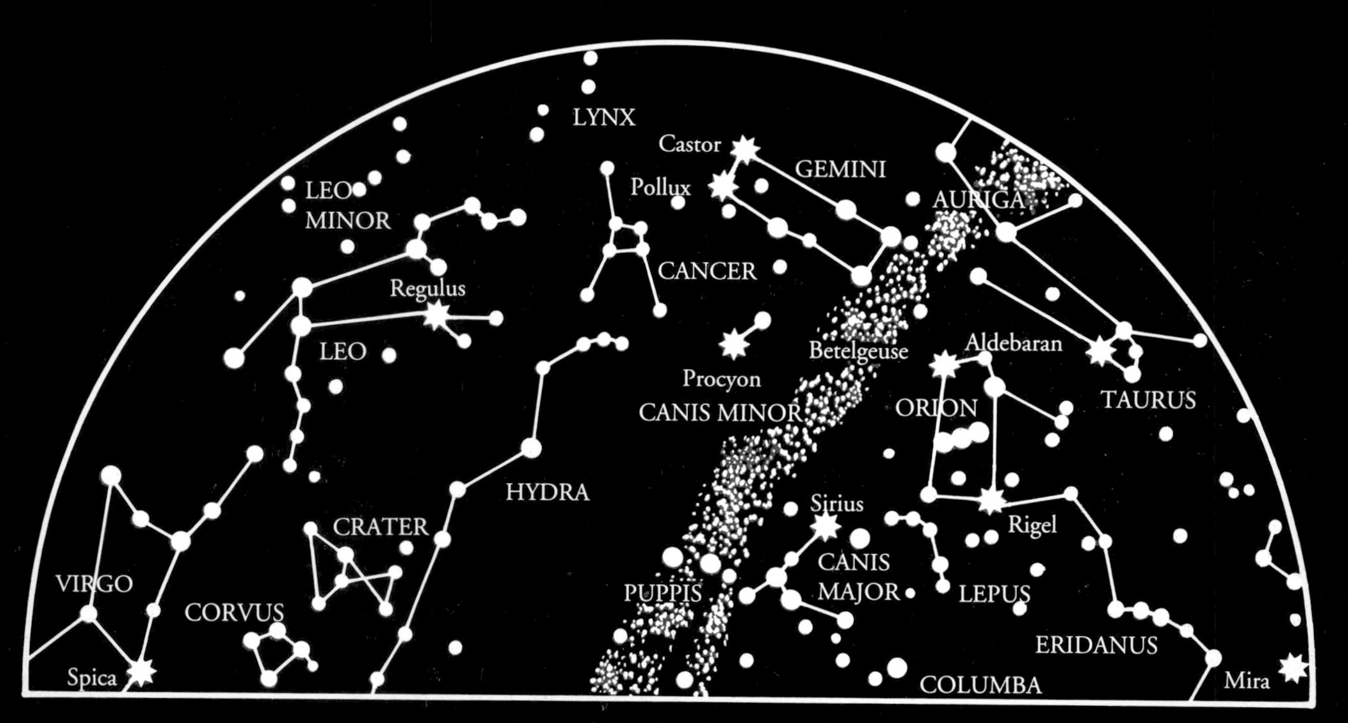

Toward the South in March

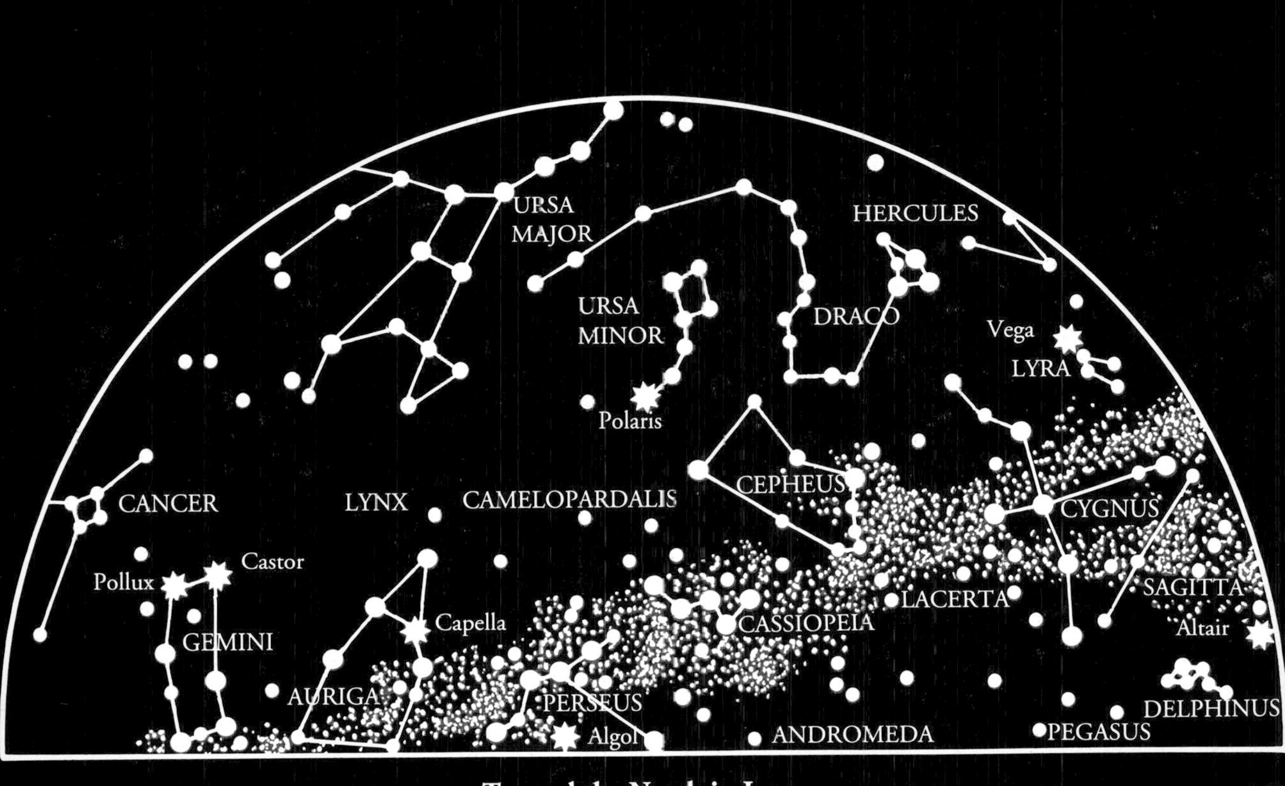

Toward the North in June

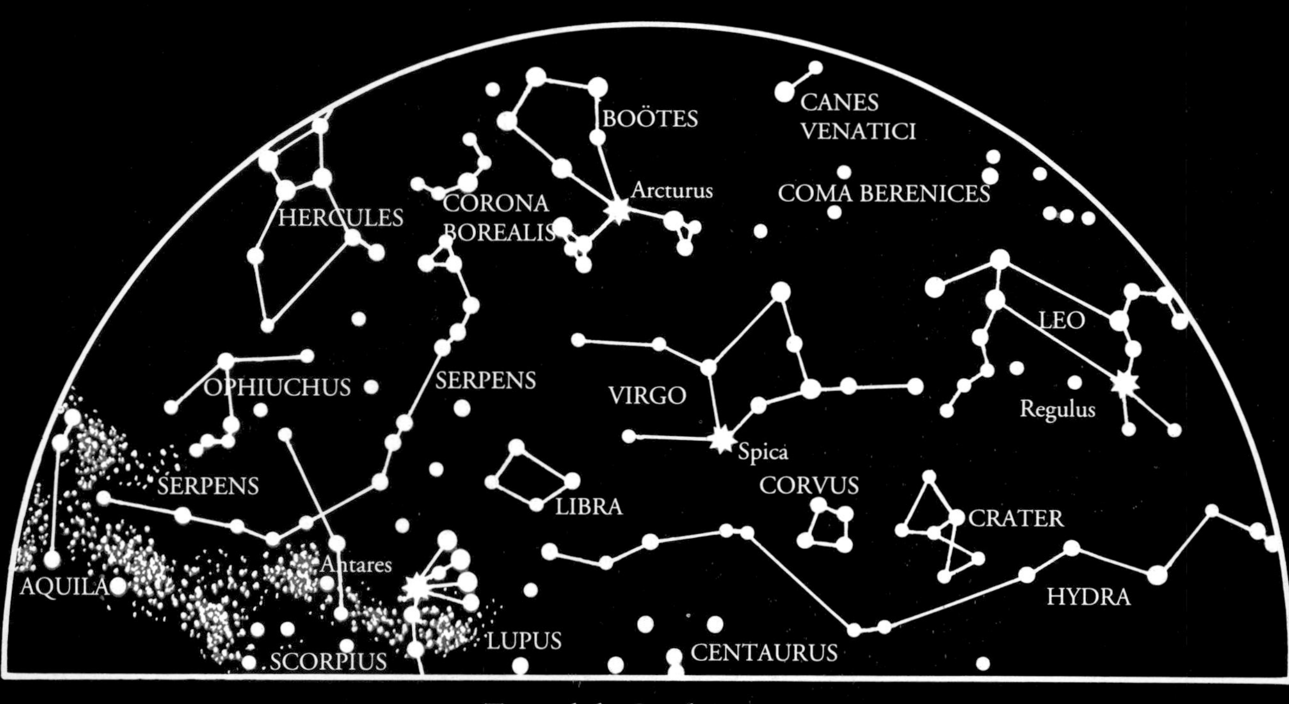

Toward the South in June

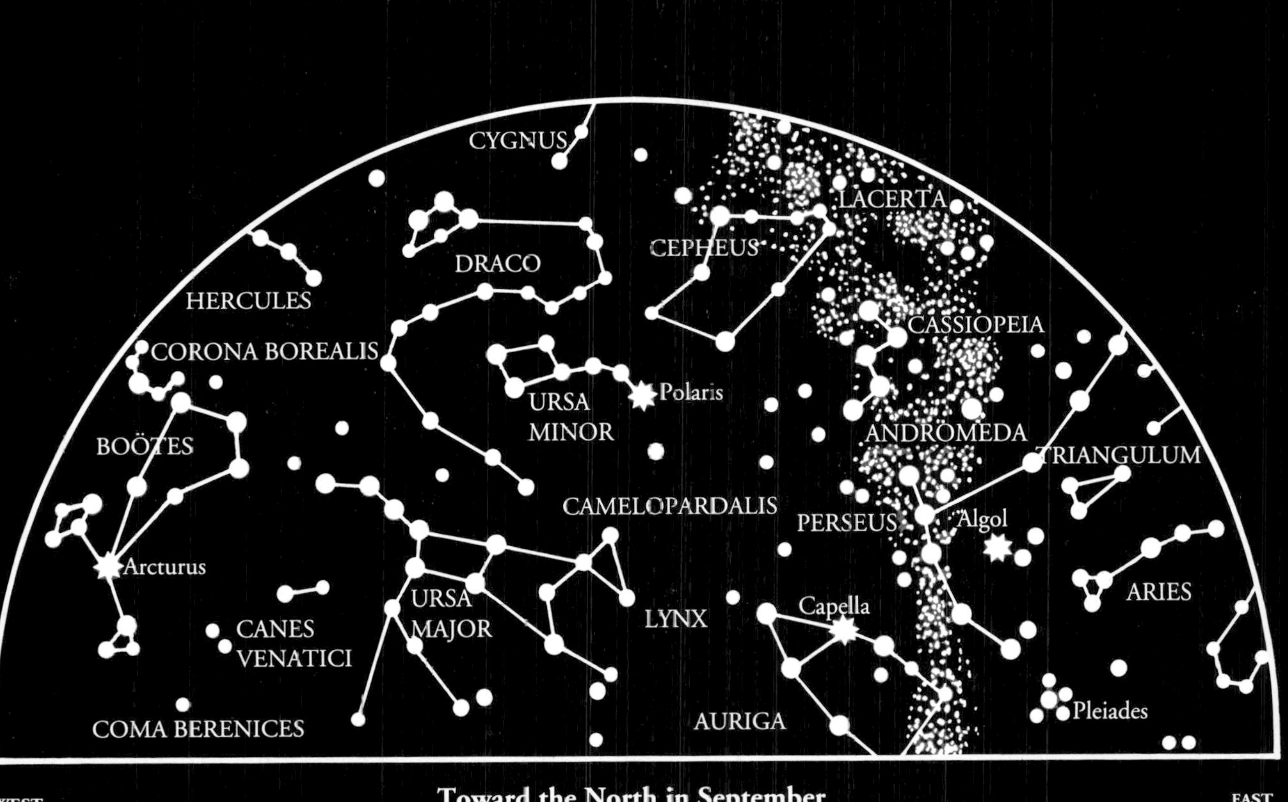

Toward the North in September

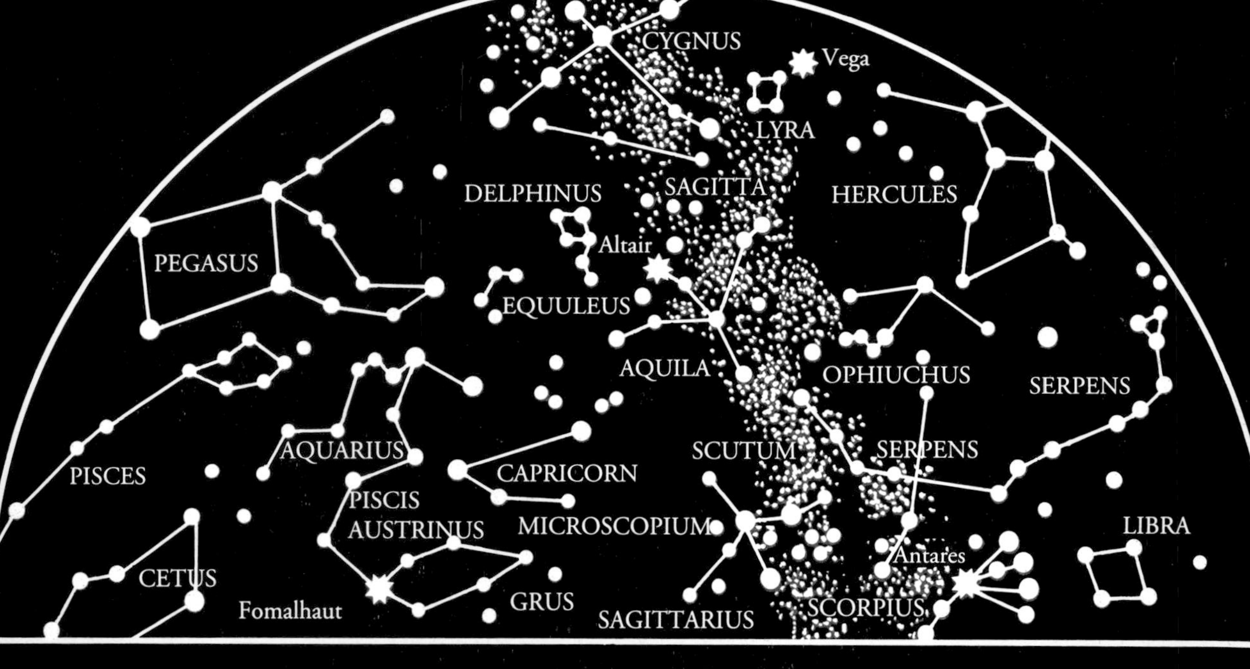

Toward the South in September

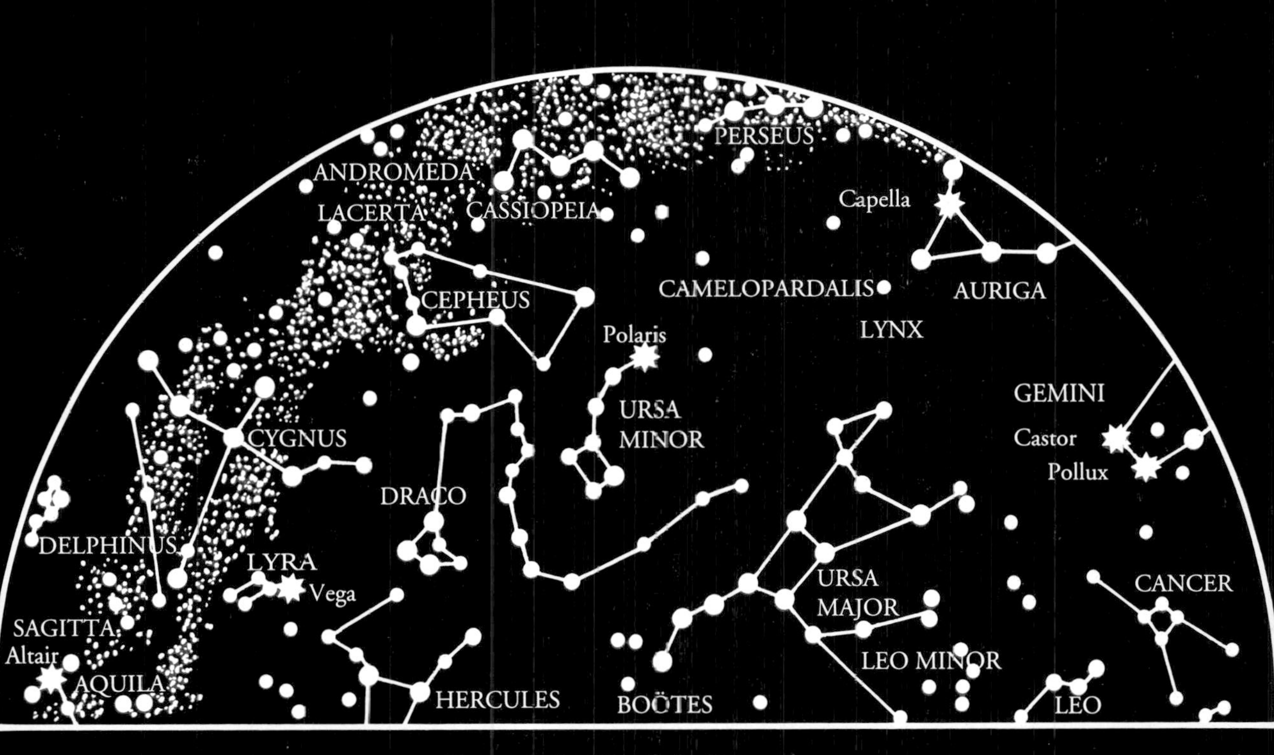

Toward the North in December

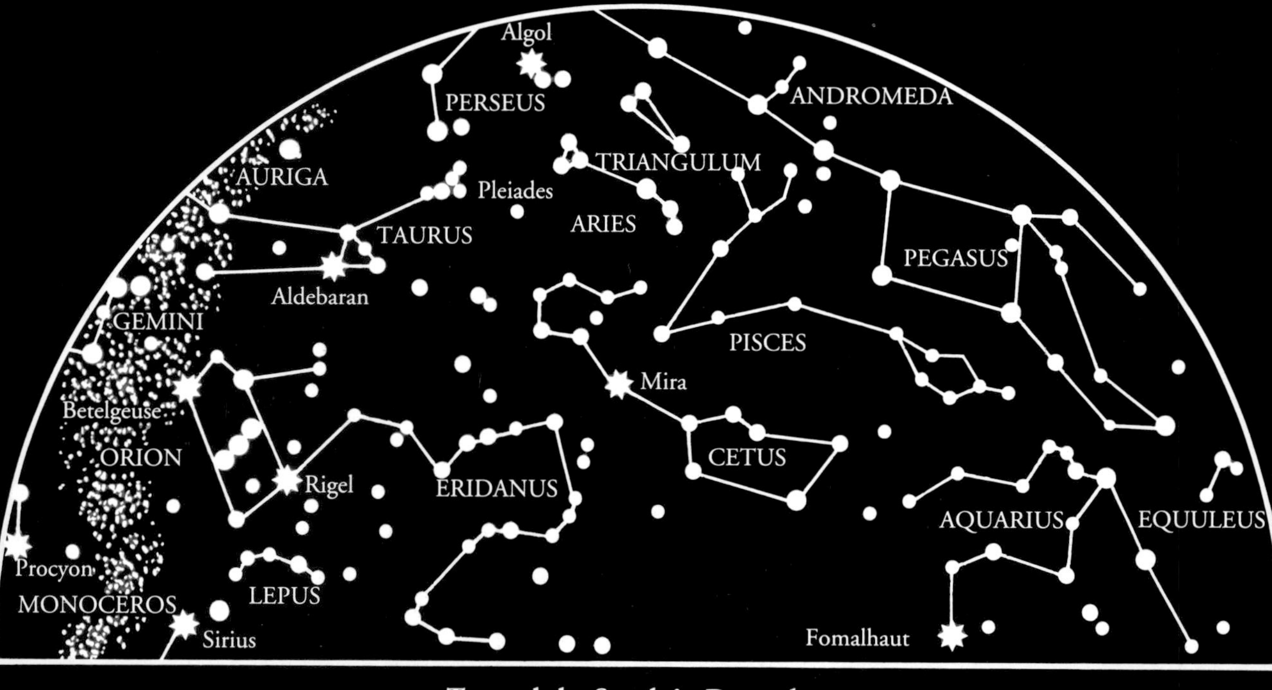

Toward the South in December

DOCUMENTS

Pierrot is a stock comic figure deriving from old French pantomines. The 19th-century toy below incorporates Pierrot and the Man in the Moon.

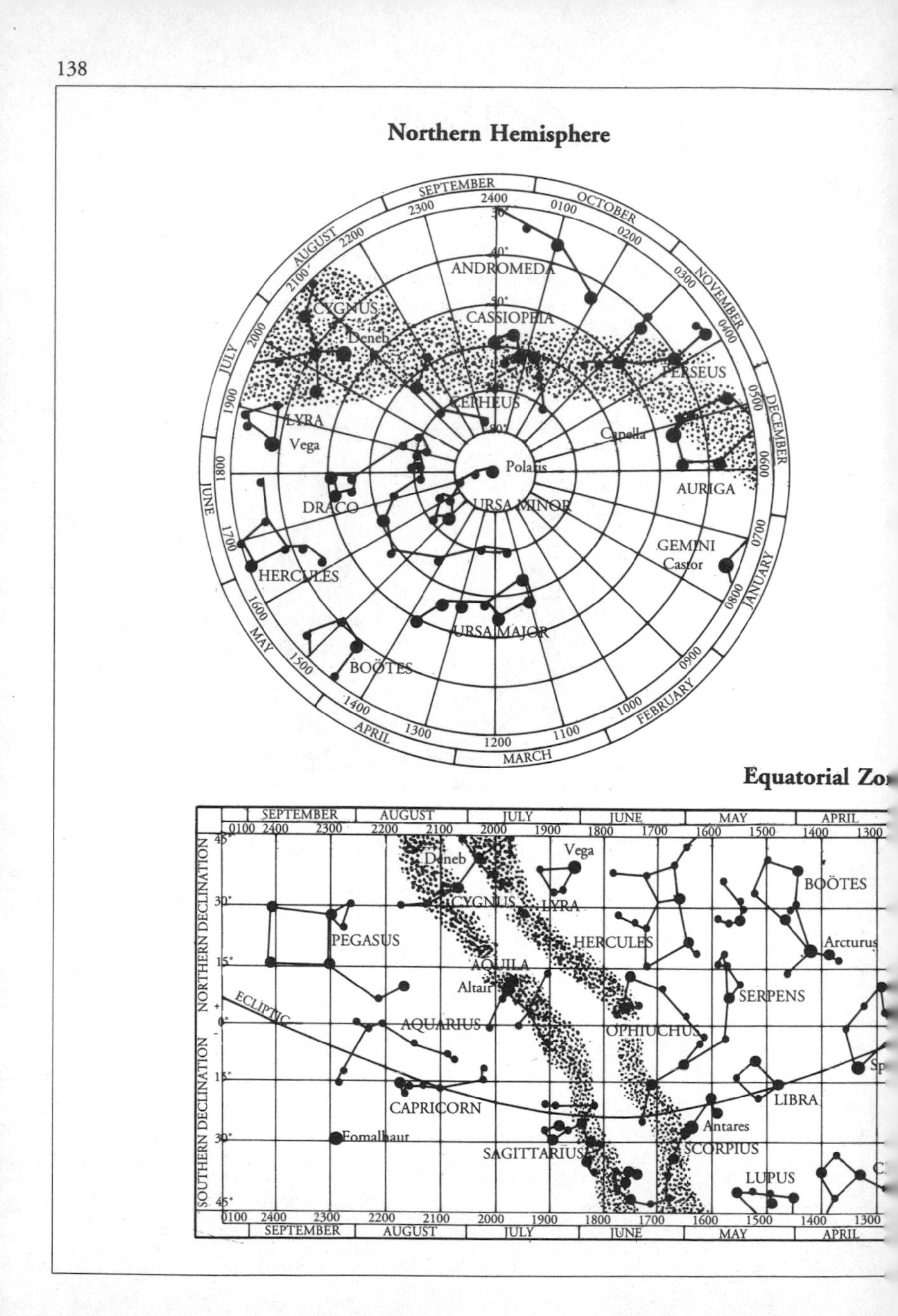
Northern Hemisphere
SEPTEMBER
OCTOBER
NOVEMBER
DECEMBER
JANUARY
FEBRUARY
MARCH
APRIL
MAY
JUNE
JULY
AUGUST
ANDROMEDA
CASSIOPEIA
CYGNUS
Deneb
PERSEUS
CEPHEUS
LYRA
Vega
Capella
Polaris
DRACO
URSA MINOR
AURIGA
GEMINI
Castor
HERCULES
URSA MAJOR
BOÖTES
Equatorial Zo
NORTHERN DECLINATION
SOUTHERN DECLINATION
ECLIPTIC
PEGASUS
CYGNUS
LYRA
HERCULES
BOÖTES
Arcturus
AQUILA
Altair
SERPENS
AQUARIUS
OPHIUCHUS
CAPRICORN
LIBRA
Antares
Fomalhaut
SAGITTARIUS
SCORPIUS
LUPUS

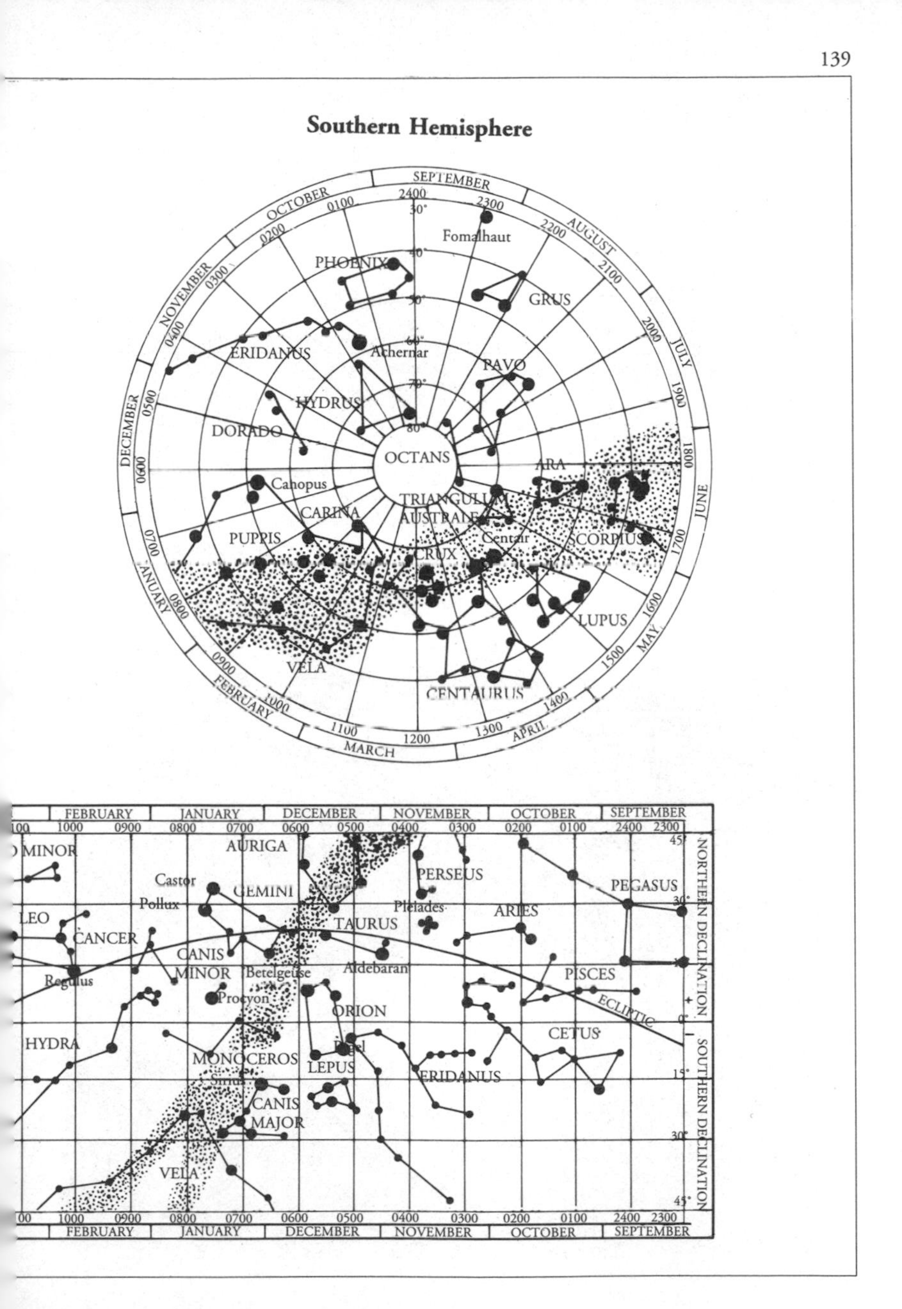
Southern Hemisphere
SEPTEMBER
OCTOBER
NOVEMBER
DECEMBER
JANUARY
FEBRUARY
MARCH
APRIL
MAY
JUNE
JULY
AUGUST
Fomalhaut
PHOENIX
GRUS
ERIDANUS
Achernar
PAVO
HYDRUS
DORADO
OCTANS
ARA
Canopus
CARINA
TRIANGULUM AUSTRALE
PUPPIS
Centaur
SCORPIUS
CRUX
LUPUS
VELA
CENTAURUS
MINOR
AURIGA
PERSEUS
Castor
GEMINI
PEGASUS
Pollux
Pleiades
ARIES
LEO
CANCER
TAURUS
CANIS MINOR
Betelgeuse
Aldebaran
PISCES
Regulus
Procyon
ORION
ECLIPTIC
HYDRA
CETUS
MONOCEROS
LEPUS
ERIDANUS
Sirius
CANIS MAJOR
VELA
NORTHERN DECLINATION
SOUTHERN DECLINATION

The Sky in All Its States

There exist as many different forms of literature about the sky as there are approaches to the study of it. Poets have written countless paeans to the sky, and some astronomers even became poets.

On 11 October 1527 a comet appeared in Vuestrie, in upper Alsace, which was part of Germany, and was visible over a prolonged period of time. Its effect was so frightening and memorable that more than a century afterward people continued to write about it.

Memorable Comet

It is quite consistent that the Comets have never appeared without foretelling some series of misfortunes, wars, plagues, famines, earthquakes, deaths, fires, floods, or other extreme devastations and calamities. Some people believe that certain constellations have bad and destructive influences and spread them to sublunar bodies,

The passing of a comet, as illustrated in a 17th-century engraving.

infecting them, bringing sterility to the Earth, followed by famine and death. Others believe that these troubles are caused by certain thick black sooty vapors, which, being combustible, rise above the middle region of the atmosphere, from where they can easily descend to a point just above the circle of the Moon, becoming thicker and condensing as they do so. As they ignite they spread a stinking and vile matter and leave marks of their terrible effects in the regions onto which they fall. But there can be no doubt that, like thunder and lightning, which are the weapons of God, there is something fatal about the Comets; the Scriptures practically tell us this: "There will be signs in the sky"; elsewhere we read that "never has one regarded" Comets with impunity.... We pass on to its horrible figure: A form of arm that emerged from a flaming cloud, dull rather than bright in color, half curved, was holding a great sword in its hand; it was bare and menacing, with its point down as though it wanted to strike or pierce not with the blade but with the point; there was a star at its tip, whose glare was terribly bright, and two other, less bright stars, appeared beside this one, one on each side of it and slightly above it. Even more terrifying was that one could see a large number of men's heads, as if they had been cut off, whose appearance was extremely hideous; they had long black beards and long hair standing on end: They appeared on both sides of this bare sword over quite a large area. What caused yet more terror was a large number of axes, picks, spears, pikes, swords, and cutlasses, all the color of blood, whose fateful appearance made one shake with fear. And indeed this comet was followed by several calamities, and the whole of Europe felt its effects intensely, as it was almost bathed in human blood. The Turks led a furious invasion of various regions, into Poland, Greece, Hungary, Persia, Arabia, and into the islands of the Archipelago. Italy suffered great devastation; several other areas, in Africa, Asia, as well as America, had their own misfortunes. [In France] there were great Wars in the South, followed by great revolutions in the Kingdoms and Provinces and by deaths, famine and plague, which are generally the companions of War. May God by his eternal providence turn all of these scourges away from our heads and may he lead the Christian Princes to a lasting peace and to an eternal alliance, which is the wish of every nation, and the rest of the Kingdoms and Provinces.

Almanac for the year 1678

Henri Pourrat's collection of French folktales provides a very nice explanation of thunder and lightning.

Thunder and Lightning

Once upon a time, the Devil played his famous trick upon Adam, our

Sixteenth-century woodcut of a comet.

grandfather, by seducing Eve and getting her to influence Adam. Envy was already tormenting him. It tormented him even more when he saw that although Adam and Eve had been expelled from paradise, they had been promised a full redemption and the chance to become the brother and sister of Our Lord.

God knows, yes God knows whether the Devil did not go to special trouble to put thistles into the cornfields and thorns onto the branches of roses.

He was proud of what he had done, but he wanted to do even better. He reflected: "Humans are fragile, more fragile than window glass, which fears blows but not noise. They, however, are even afraid of an uproar. I have to invent something that strikes, that rumbles, that shatters rocks and oaks. Since I will not be able to reduce everything to ashes, I want, at least, to strike everything with terror. It will have to involve such a racket that every creature will tremble from one edge of the sky to the other!"

The Devil went to find God:

"I shall make thunder: I shall terrify your children."

And God immediately had the idea of a light that would dazzle them. It would thus remind them of the light before the terror struck:

"I shall make lightning: they will see it first of all: They will entrust themselves to me."

Henri Pourrat
French Folktales, 1989

Solar eclipses inspired two astronomers from different centuries: the Italian Ruggiero Giuseppe Boscovich in the 18th century and Camille Flammarion of France in the 19th.

When Phoebus Robs the Earth of the Fire of His Rays

Why is it that when all clouds have been driven from the sky and when Phoebus [Apollo] pours his rays onto the Earth and shines triumphantly, when nothing in the air can diminish the force of his fire, why does thick darkness sometimes suddenly cover over the dazzling face of this God of the Day? Why does impatient Night, anticipating the hour of her domain, come in the middle of a fine day and spread her dark veil and allow the astonished eyes of mortals to see only the weak glow of the stars? And why, at the very moment at which she was celebrating her reign in the sky, does Phoebe either retreat into the heart of darkness or show us the sad spectacle of her reddened and bloodied face? These are the phenomena that my Muse undertakes to celebrate and whose causes are explained by my lines.

You who also reign over ethereal Olympus, over the cherished Olympus of the nine Sisters, divine Phoebus, reveal to my eyes the secrets of Nature, penetrate my heart with your sacred rays. Whether I am telling mortals how you deprive the Earth of the fire of your rays, or am showing them how you refuse to share them with Phoebe your sister, it is your interest that inspires me, it is you that I am celebrating. May my verses be worthy of the God I am praising.

And you, who are, for me, the most interesting and most beloved of the nine Sisters, you who, despising earthly regions, raise your chariot to the skies and hide your face amid the stars, divine Urania, favorable to my wishes, support and never abandon a like those golden nails that shine attached to our paneling, these various stars are also fixed to the vault of the heavens at the same height. Carried through the void or through a subtle air that is dispersed in the vastness of space, they rise unevenly toward Olympus and leave between themselves and the Earth varying distances.

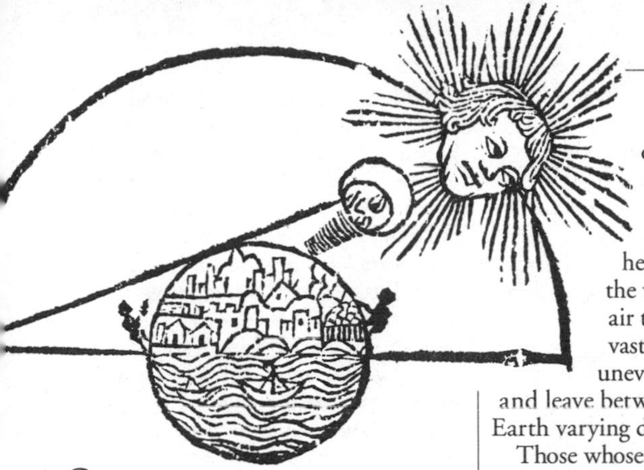

Those whose trembling light constantly flickers, those whose fine rays only strike our eyes as an extremely sharp line, those stars always present us with the same respective position; a similar appearance always reveals them to be equally far apart: They have therefore been called fixed stars. Placed at immense distances, on the edges of the vast Universe, their height does not allow a single one of them to be analyzed by the onlooker. Their fire may well equal, or even go beyond that of Phoebus; but their rays, scattered in the air and worn out by such a long journey, can barely overcome the shadows of the night. If you were to be transported by a bold flight into the loftiest regions of the sky, you would see the sun itself grow gradually smaller and be finally plunged into the darkness of night.

Ruggiero Giuseppe Boscovich
The Eclipses
(poem in six songs, dedicated to His Majesty Louis XVI of France, 1779)

Sixteenth-century engraving of an eclipse.

poet who is devoted to you.

But rather than to Phoebus and to the learned Sisters, it is to you...that I address my wishes. Your genius presides over these learned assemblies, at which the fathers of the sciences and of the arts penetrate the most hidden secrets of nature. It is under your auspices that, from the banks of the Thames, they spread light over the vast regions of the Universe. Support with them a mortal who is enriched by their works and by yours. Yes, it is to you above all, illustrious Sages, that my Muse owes the object of its songs....

Anxious to know the different eclipses of the sun and of the moon, to learn of their secret causes, your first concern will be to acquire a perfect knowledge of the celestial region; you will study the position of the stars and their movements.

First, as you contemplate the sky and see the countless stars that a fine night reveals to you, or a languishing Phoebe contracting her luminous disk, or the Sun himself raising himself onto the banks of dawn, do not believe that,

When the Star of the Day Disappears at High Noon

The last *total* eclipse of the sun visible in France was that of 8 July 1842, seen as partial at Paris but total in the South of France. I admit that I was not an eyewitness to it, first because I did not live in the zone of central eclipse, second, and especially, on account of my extreme youth [the author was then four months and eleven days old]; but he who was afterward by his noble and powerful writings my master, François Arago, had returned to the Eastern Pyrenees, his birthplace, expressly to observe it, and the following is an extract from his account of the sight:

"The hour for the beginning of the eclipse approached. Nearly twenty thousand persons, with smoked glasses in hand, examined the radiant globe projected on an azure sky. Scarcely had we, armed with our powerful telescopes, begun to perceive a small indentation on the western limb of the sun, when a great cry, a mingling of twenty thousand different cries, informed us that we had anticipated only by some seconds the observation made with the naked eye by twenty thousand unprepared astronomers. A lively curiosity, emulation, and a desire not to be forestalled would seem to have given to their natural sight unusual penetration and power.

"Between this moment and those that preceded by very

little the total disappearance of the sun we did not remark in the countenances of many of the spectators anything that deserves to be related. But when the sun, reduced to a narrow thread, commenced to throw on our horizon a much-enfeebled light, a sort of uneasiness took possession of everyone. Each felt the need of communicating his impressions to those who surrounded him: hence a murmuring sound like that of a distant sea after a storm. The noise became louder as the solar crescent was reduced. The crescent at last disappeared, darkness suddenly succeeded the light, and an absolute silence marked this phase of the eclipse so that we clearly heard the pendulum of our astronomical clock. The phenomenon in its magnificence triumphed over the petulance of youth, over the levity that certain men take as a sign of superiority, over the noisy indifference of which soldiers usually make profession. A profound calm reigned in the air; the birds sang no more.

"After a solemn waiting of about two minutes, transports of joy, frantic applause, saluted with the same accord, the same spontaneity the reappearance of the first solar rays. A melancholy contemplation, produced by unaccountable feelings, was succeeded by a real and lively satisfaction of which no one thought of checking or moderating the enthusiasm. For the majority of the public the phenomenon was at an end. The other phases of the eclipse had hardly any attentive spectators, apart from those devoted to the study of astronomy."

Camille Flammarion
Popular Astronomy, 1894

Observing an eclipse in Paris, 17 April 1912.

When the upholders of justice involve themselves with things that fall from the sky, the situation can border on the ridiculous. The following story illustrates this phenomenon very well.

The landing of a meteorite near an Italian village in 1937 proved that meteorites do not always fall in deserted areas.

The Trial of the Meteorite

On 12 June 1895 I traveled to the Château de Grammont where a Monsieur François Douillard was waiting for me. Monsieur Douillard was the farmer near whom a meteorite had fallen fifty-four years earlier and who had been its first owner.

Douillard, seventy-seven years old at the time when I saw him, is a short man, in good health and very alert. He told me that while still at work an hour after sunset, he had heard, coming with amazing speed from the direction of Legé, that is, from the west, a terrifying whistling, followed by a huge explosion and by a thud of something falling 100 or 150 yards away from him. According to Douillard there had been no luminous trail and the impact had been heard at Lucs....

It had hit the ground at the bottom of a furrow separating two beds of vines, one of which belonged to Mrs. Guichet from La Bernardière, the other to Mr. Vollard of Legé, and was lying near a crater [about a foot] deep, which it had formed in its fall but outside of which it had come to rest.

François Douillard took the meteorite that had given him such a fright and sold it to Dr. Mercier, the owner of the nearby Château de Grammont....

Disputes soon arose concerning the ownership of the meteorite acquired by Dr. Mercier. Mr. Vollard and Mrs. Guichet asserted their rights to this meteor, which in falling had by chance touched the borders of their respective properties. When their negotiations could not be resolved, they decided to go to court, and it was Mr. Vollard who sued Dr. Mercier. A judgment was given by the tribunal of La Roche-sur-Yon, called at that time Bourbon-Vendée.

Because of the unusual nature of the dispute that this meteorite had provoked, I think it is interesting to reproduce a portion of the text from that judgment....

"Considering the fact that the stone that the case concerns is an aerolite that before its fall to Earth was obviously nobody's property, and that Vollard, who does not pretend he really ever owned it, claims it be granted to him as the owner of the parcel of land that the stone touched when falling and on which it came to rest;

"Considering that Mercier, without recognizing this last fact, which, on the contrary, he denies, claiming it to be irrelevant to the case since the stone, before its fall, had not belonged to anyone, it should, according to him, belong to the first occupant or to the inventor, in whose place he puts himself, as has just been said;

"Considering that in law, which our present legal system recognizes as Roman law, the existence of things that have never had an owner or whose owner is not known;

"Considering that, among these things, the greatest number is subject to a private property; that one must therefore find *a priori* to whom that property should be attributed, except when a lost object is involved, and its restoration to the first owner who makes himself known within the time limits that have been set;...

"Considering that the only plausible exception by which one can legally challenge, in the interests of an individual, the right of the first occupier, is that resulting from the right of accession, as a consequence of which one asserts, as does Vollard, that the vacant object that does not have an owner becomes the property of the person on whose land it is found;

"Considering that in order to judge the merit of this exception one must agree on the value of the terms used by the law-giver when he said that the right of accession 'is the right of the owner over all that unites with and is incorporated into the thing';...

"Considering that these conditions, which appear essential to form the right of accession, are by no means found in this case; since one cannot say that the aerolite in question had either been united with and incorporated into Vollard's field in such a way as to form part of it or had raised its intrinsic value or that of its periodic products;

"Considering that in fact one must recognize that stones from quarries or elsewhere that are found in a field are accessories of it, because these stones, forming part of the Earth, with which and for which they have been created, from their very origin, also and as a consequence, form part of the fields on which they have been placed; but that one cannot say the same about the aerolite, which is of a quite different order and is foreign to the Earth, where it only arrived as a result of an accident that propelled it from its place of origin; that this aerolite cannot be identified with the ground onto which it fell any more than could a watch or any other precious or nonprecious object that a traveler might have lost there and nobody ever asserted that an object of such a nature was united by virtue of accession to the field on which it was found;...

"Considering, indeed, that it is impossible reasonably to assimilate it to the violation of a home or to the clandestine introduction into a citizen's house, the act, which is quite innocent in and of itself, of entering into one of his unenclosed fields and outside his residence, while he had not manifested his intention of opposing it;

"For these reasons the Tribunal declares Vollard's claim to be without foundation."

M. A. Lacroix
The Meteorite of Saint-Christophe-la-Chartreuse

The flaming light effect known as Saint Elmo's fire takes its name from the Italian bishop and patron saint of sailors. Sometimes seen in stormy weather, these electrical discharges have been associated with ephemeral stars that appear over land and sea. Among sailors, especially, many legends grew up around the phenomenon. The following anecdote is taken from the journal of a 19th-century sailor.

Saint Elmo, Pray for Us

During a stormy night one noticed on board fires that played at each end of our main yard. This bright and blue flame, like those that one lights on the punch that is served in cafés, aroused my curiosity for the first time.

"What on earth is that?" I asked a sailor in amazement.

—"Saint Elmo's fire, sir."

—"Ah, yes, it burns!"

—"One had better say that it is the sailor's friend. Do you see that kind of flame? Well, if the officer of the watch told me: 'Climb up by yourself, pull down the small topsail' [which is in fact quite heavy for just one man], I would pull it down on the double because that fire would go up with me up the rigging to help me, as it helps all sailors."

—"But how can you take such a story seriously? It is quite simply, as I remember having read, a natural effect, an electrical discharge that, like a fluid of this kind, seeks points."

—"How can I believe that story? An effect of *lubricity, electrical* discharge, as you please. But it is no less true that that fire, which resembles a glass of brandy that is alight, is the soul of a poor sailor who drowned in the sea in a storm. So you see, when the weather is about to get worse, the soul of the sailors who have taken one drink too many from the great pond, comes and warns their comrades that a dangerous storm is approaching."

—"My word, just in case it is true I want to see if I can touch the soul of a dead man, and I shall go straight to the running board of the yardarm to catch up with your Saint Elmo's fire."

I climbed to the end of the yardarm, as I had said I would, to the great surprise of my companion, who saw a kind of profanation in the intention that I had of going needlessly to bother what he called the sailor's friend.

As my hand gradually got closer to the Saint Elmo's fire, the fluid moved up and down and away and did not come back until I had withdrawn it. This sort of little war between it and myself greatly amused the men of the watch, and they said to me again and again:

—"Oh, that one is meaner than you or us."

A sailor from Lower Brittany cried out to me:

—"Do you want me to make it disappear?"

"Yes," I replied.

And he made the sign of the cross. The fire did indeed vanish at that very moment, and that instantaneous coincidence between the fire's disappearance and the sign of the cross made by the devout man helped to engrave a superstition even more profoundly in the imagination of those good people.

Edouard Corbière
The Slave Trader, 1832

Saint Elmo's fire as interpreted in a 19th-century engraving by Gustave Doré.

Celestial Liturgy

The Apocalypse of, or the Revelation to, Saint John, which makes up the last book of the New Testament, outlines revelations made by God to the people. It is remarkable for the strange character of the visions it contains, the elaborate symbolism that expresses them, and the dramatic scenes of the end of the world that it evokes.

In the following apocalyptic vision the seer rises to heaven to look upon a series of grandiose scenes. Within the setting of a celestial liturgy appears the lamb, who takes possession of the divine plans and breaks open their seven seals one by one. The opening of the seventh seal signals the punishment of all evildoers of the world. Seven trumpets underline the various forms that punishment will take.

When the Lamb opened the seventh seal, there was silence in heaven for about half an hour. And I saw the seven angels who stand before God, and seven trumpets were given to them.

In the engraving below the Virgin dictates the Apocalypse to Saint John.

Another angel with a golden censer came and stood at the altar; he was given a great quantity of incense to offer with the prayers of all the saints on the golden altar that is before the throne. And the smoke of the incense, with the prayers of the saints, rose before God from the hand of the angel.

Then the angel took the censer and filled it with fire from the altar and threw it on the Earth; and there were peals of thunder, rumblings, flashes of lightning, and an earthquake.

Now the seven angels who had the seven trumpets made ready to blow them.

The first angel blew his trumpet, and there came hail and fire, mixed with blood, and they were hurled to the Earth; and a third of the Earth was burned up, and a third of the trees were burned up, and all green grass was burned up.

The second angel blew his trumpet, and something like a great mountain, burning with fire, was thrown into the sea. A third of the sea became blood, a third of the living creatures in the sea died, and a third of the ships were destroyed.

The third angel blew his trumpet, and a great star fell from heaven, blazing like a torch, and it fell on a third of the rivers and on the springs of water.

The name of the star is Wormwood. A third of the waters became wormwood, and many died from the water, because it was made bitter.

The fourth angel blew his trumpet, and a third of the sun was struck, and a third of the moon, and a third of the stars, so that a third of their light was darkened; a third of the day was kept from shining, and likewise the night.

Then I looked, and I heard an eagle crying with a loud voice as it flew in midheaven, "Woe, woe, woe to the inhabitants of the Earth, at the blasts of the other trumpets that the three angels are about to blow!"

And the fifth angel blew his trumpet, and I saw a star that had fallen from heaven to Earth, and he was given the key to the shaft of the bottomless pit; he opened the shaft of the bottomless pit, and from the shaft rose smoke like the smoke of a great furnace, and the sun and the air were darkened with the smoke from the shaft.

Then from the smoke came locusts on the Earth, and they were given authority like the authority of scorpions of the Earth. They were told not to damage the grass of the Earth or any green growth or any tree, but only those people who do not have the seal of God on their foreheads. They were allowed to torture them for five months, but not to kill them, and their torture was like the torture of a scorpion when it stings someone. And in those days people will seek death but will not find it; they will long to die, but death will flee from them.

In appearance the locusts were like horses equipped for battle. On their heads were what looked like crowns of gold; their faces were like human faces, their hair like women's hair, and their teeth like lions' teeth; they had scales like iron breastplates, and the noise of their wings was like the noise of many chariots with horses rushing into battle. They have tails like scorpions, with stingers, and in their tails is their power to harm people for five months.

They have as king over them the angel of the bottomless pit; his name in

Hebrew is Abaddon, and in Greek he is called Apollyon.

The first woe has passed. There are still two woes to come.

Then the sixth angel blew his trumpet, and I heard a voice from the four horns of the golden altar before God, saying to the sixth angel who had the trumpet, "Release the four angels who are bound at the great river Euphrates."

So the four angels were released, who had been held ready for the hour, the day, the month, and the year, to kill a third of humankind. The number of the troops of cavalry was two million; I heard their number.

And this was how I saw the horses in my vision: the riders wore breastplates the color of fire and of sapphire and of sulfur; the heads of the horses were like lions' heads, and fire and smoke and sulfur came out of their mouths. By these three plagues a third of humankind was killed, by the fire and smoke and sulfur coming out of their mouths. For the power of the horses is in their mouths and in their tails; their tails are like serpents, having heads; and with them they inflict harm.

The rest of humankind, who were not killed by these plagues, did not repent of the works of their hands or give up worshiping demons and idols of gold and silver and bronze and stone and wood, which cannot see or hear or walk. And they did not repent of their murders or their sorceries or their fornication or their thefts.

And I saw another mighty angel coming down from heaven, wrapped in a cloud, with a rainbow over his head; his face was like the sun, and his legs like pillars of fire. He held a little scroll open in his hand. Setting his right foot on the sea and his left foot on the land, he gave a great shout, like a lion roaring.

And when he shouted, the seven thunders sounded. And when the seven thunders had sounded, I was about to write, but I heard a voice from heaven saying, "Seal up what the seven thunders have said, and do not write it down."

Then the angel whom I saw standing on the sea and the land raised his right hand to heaven and swore by him who lives forever and ever, who created heaven and what is in it, the Earth and what is in it, and the sea and what is in it:

"There will be no more delay, but in the days when the seventh angel is to blow his trumpet, the mystery of God will be fulfilled, as he announced to his servants the prophets."

Then the voice that I had heard from heaven spoke to me again, saying, "Go, take the scroll that is open in the hand of the angel who is standing on the sea and on the land."

So I went to the angel and told him to give me the little scroll; and he said to me, "Take it, and eat; it will be bitter to your stomach, but sweet as honey in your mouth." So I took the little scroll from the hand of the angel and ate it; it was sweet as honey in my mouth, but when I had eaten it, my stomach was made bitter.

Then they said to me, "You must prophesy again about many peoples and nations and languages and kings."

The Apocalypse of Saint John
also known as Revelation
8, 9, 10

This late-15th-century woodcut is from a series of Apocalypse scenes by the German artist Albrecht Dürer.

The Republican Calendar

Calendars divide time into periods of varying length to suit social customs and correspond to natural phenomena, most of which are astronomical in nature. The most obvious phenomenon by which to measure time in reasonable periods is the lunar month, and over the years many lunar calendars were created. In the long term, however, the organization of the solar year, which regulates the seasons, was vital for agriculture. Yet because neither the lunar month nor the solar year has an even number of days, just as the solar year is not made up of an even number of lunar months, there arose many difficulties in marking out consistent, and universal, divisions of time.

"The era was both historical and astronomical," said historian Jules Michelet of the period surrounding the French Revolution, which began in 1789. The desire for radical change born of that movement even led the revolutionary Convention to adopt a new calendar, which was used in France from 1793 until 1806, when it was replaced by the Gregorian calendar already adopted by the United States and Great Britain and still in use today.

A Revolutionary Calendar for an Age of Revolution

A law passed by the Convention on 6 October 1793 put the new calendar

B*rumaire,* the second month of the republican calendar, from a contemporary engraving (below). At right are two illustrations from the second year of the calendar.

into effect. The proclamation of the Republic had taken place on 22 September 1792, the day of the autumnal equinox. The revolutionaries made use of this coincidence and took this date as the beginning of the new era, declaring the day on which the autumnal equinox occurs at the Paris meridian to be the first day of the new year.

The year is made up of twelve months, each lasting thirty days. The names of the months, thought up by Fabre d'Eglantine, a member of the Convention, have a wonderful ring to them and are extremely poetic. Their endings correspond to the seasons:

—Autumn: *Vendémiaire* [harvest month], *Brumaire* [misty month], *Frimière* [frosty month]

—Winter: *Nivôse* [snowy month], *Pluviôse* [rainy month], *Ventôse* [windy month]

—Spring: *Germinal* [germinating month], *Floréal* [flowering month], *Prairial* [grassy month]

—Summer: *Messidor* [month of reaping], *Thermidor* [hot month], *Fructidor* [fruitful month]

Etymologists have criticized these magnificent names, but a more serious objection may be made to them: The members of the Convention had hoped that their calendar, like the metric system, would be adopted by every nation; in fact, the names that they invented were appropriate only to the climate to be found at the time in France.

The days of the month were arranged into three periods of ten days,... and the names themselves correspond numerically to their position within these; they were: *primidi, duodi, tridi, quartidi, quintidi, sextidi, septidi, octidi, nonidi,* and *decadi.*

The year closed with five complementary days, placed after *Fructidor.* A sixth day, called the day of the Revolution, appeared every fourth year and was called the sextile. Sextile years did not coincide with the Gregorian bisextile years (they were years 3, 7, and 11).

The calendar of the French republic. The twelve months, whose names were derived from aspects of the climate or from the harvests, were divided among the seasons (here beginning with autumn) in groups of three.

The republican calendar lasted thirteen years. By a decree of 9 September 1805, Napoleon declared it abolished as of 1 January 1806. Because the year 1 was never used, as the calendar was instituted on 15 *Vendémiaire* of the year 2, there were only twelve years in which it was applied: No authentic act

In the republican calendar the year begins on the day on which the autumnal equinox falls in Paris. Astronomers were asked to determine the moment at which the phenomenon occurred, and a decree subsequently fixed the start of the year.

One difficulty had to be resolved: When the sun was at the autumnal equinox very close to midnight in Paris, the resulting tiny gap of time led to an uncertainty of a whole day. According to the astronomer Jean Baptiste Joseph Delambre, this was would happen in the year 144 of the Republic.

By causing the beginning of the year to depend on a calculation relating to the Paris meridian, the founders of the republican calendar, who hoped to see their invention universally adopted, committed a serious psychological error: The force of nationalities expresses itself even in such matters, and the adopting of an international system of longitudes and coherent time zones faced many difficulties, which were not resolved until the twentieth century. Objections to the names of the months added to these difficulties and condemned the system to a strictly local usage.

Could it even be said that the calender found any favor within France? Obviously not. Legislators had underestimated the strength of the links with the past, the endurance of anniversaries in people's hearts: The new chronology was a break with history. It may have made sense for the future but it ignored the past, even the most recent past. It did not have the time to put down roots in the soul of the people and disappeared without regret, even from contemporaries....

Paul Couderc
The Calendar, 1946

could bear the date of year 1. Year 14, which began on 23 September 1805 (Gregorian), lasted only for three months and eight days.

One finds in every *Annual of the Office of Longitudes* (especially in odd years) a table of concordance between the republican and Gregorian dates.

Chapbooks

Chapbooks were collections of poems, songs, legends, and religious tracts. They were generally small and were distributed in large numbers, often by itinerant booksellers. For a time almanacs were grouped within this category of popular literature. Often feared as a means of disseminating propaganda to the masses, these little books almost completely disappeared in many areas during the 19th century as they fell victim to censorship commissions and increasing controls placed upon the publication and distribution of printed materials.

The Almanacs

The etymology of the word *almanac* is uncertain. It may be of Arabic origin (*manach* means a count); or of Saxon origin (*almooned* is a reference to a system of classifying the lunar months); or even of Celtic origin, since *almanac* could also refer to a monk who was a prophet of the 3rd century AD.

It is clear that the almanac has since its inception been linked to the classification or interpretation of time. It was probably one of the most ancient forms of books. We know that the Chinese, the Egyptians, and the Greeks used almanacs; manuscripts dating from long before the age of printing have come down to us. In the 15th century, the famous *Shepherd's Calendar,* which became the preferred model for the almanac, appeared. At the same time one finds traces of corporate almanacs. But it was not until the 17th century that the almanac made its true appearance in popular literature. At that time the contents of the booklet, which was to enjoy a great popularity among rural populations, was based above all on astrological and prophetic considerations: The weather was predicted according to the method of astrometeorology.

In fact it was astrology that played the predominant role in the division and arrangement of time since it was believed that the stars, by their influence on the seasons, the plant cycle, animal life, and human behavior, determine the order of the universe, of the macrocosm and the microcosm.

That is why so often we see represented on the title pages of the almanacs that all-powerful character the astrologer: He knew the secrets of

The Print Seller. Engraving after a drawing by Jules David.

nature, he could predict the weather, and he could prophesy future events.

This explains the popularity of the collections of prognostications such as those of Thomas-Joseph Moult or the famous prophecies of Nostradamus. While these two men did in fact exist, other astrologers were very often the inventions of publishers who specialized in chapbook literature.

In the 18th century we notice in the production of almanacs a very clear tendency, marked by the influence of the Enlightenment thinkers, to denounce astrology and to attribute a greater role to divine will alone. The utilitarian and encyclopedic aspect of the almanac became more prominent and it began to be a subject for entertainment or amusement.

During the French Revolution, the method of distribution of the almanac, which could reach the most isolated villages by means of the traveling bookseller, was found to be useful in spreading the new ideology. In the 19th century the almanac showed the existence of a renewed interest in astrology, as is proved by the popularity of the Mathieu Laensbergh almanacs, published in Lille, Rouen, and Liège.

Some very well known titles still have not completely disappeared even today. *The Limping Messenger*, the first edition of which dates from the beginning of the 18th century, gave rise to a number of similar publications whose popularity among the rural population of eastern France has by no means declined, as is shown by the *Almanac of the Farmers of the Upper Rhine*, which continues to appear today, or even *The Limping Messenger* itself, which is more widely sold in Switzerland.

After the Rain Comes the Good Weather: Meteorology
Musée National des Arts et Traditions Populaires

In its most popular form the almanac was aimed above all at the shepherd and farmer, as exemplified by The Great Calendar and Compost of the Shepherds, *which became a model for subsequent almanacs.*

The Great Calendar of the Shepherds

First published in the late 15th century, *The Great Calendar and Compost of the Shepherds* is one of the great successes of chapbook literature; in three centuries there were no fewer than forty editions of it in the French language. The first editions were distinguished by the quality of their woodcut engravings. It was not until 1657 that the first truly popular edition of this almanac appeared, published in Troyes, by Nicolas Oudot. Also known as the *Shepherd's Calendar*, its immense popularity was such that all the great printers who specialized in the production of chapbooks published at least one edition of the work.

Perpetual Prognostication for Farmers

In Order to Know What the Weather Will Be:

If the sky is red in the morning it means that it will rain in the evening, but when it is red in the evening it foretells good weather for the next day.

If when the Sun rises in the morning there are long rays going down to the ground through the clouds, it means that they are drawing in water and that the fine weather will not last.

Woodcuts from a 16th-century edition of the *Shepherd's Calendar* (below) and from the *Almanac of the Good Farmer*, 1792 (above right).

In Order to Know What the Weather Will Be By Looking at the Moon:

If the Moon is blue it means the weather will be rainy; if it is red it means there will be winds; if it is white it means fine weather.

On the New Moon:

Note that if there is good weather the Tuesday immediately following the new moon, there will be two fine days. If it is rainy the day will probably be wet.

On White Frost:

Note that for as many white frosts as occur before Saint Michael's Day, and for as many days afterward, there will be an equal number of white frosts before Saint George's Day and afterward.

On Winter:

Take the breast of a duck in Autumn or afterward and have a good look at it: If it is entirely white it means that we shall have a short winter; if it is red and then white it means that we shall have the winter at the start of the season; and if it is white at the front end and red in the middle it means that we shall have cold weather in the middle of winter; and if it is red toward the very back it means that the winter will come late.

In Order to Know What the Weather Will Be in Every Week of the Year:

In order to arrange their work for the whole of each week, the old farmers notice what the weather is like on Sunday from about 7 to 10 AM; if it rains during that time most of the following week will be rainy.

Another Rule for Knowing What Weather There Will Be:

If a rainbow appears in the east as the sun is setting, there will be thunder or rain, according to the nature of the moon, and that is even more certain if two or more are seen together. If one is seen in the evening it means there will be fine weather the next morning. When several rainbows are fixed in the sky, appearing perfect and for a long period of time, it signals prosperity and peace; if [a rainbow] changes quickly and early there probably will be disturbances and storms, something that can apply to the rainbows that the Moon sometimes creates as well as to those formed by the Sun.

Master Antoine Maginu
Perpetual Prognostication of the Farmers, 1736

Although The Mirror of Astrology, *a text dating from the end of the 16th century, was never reprinted in the almanacs, it probably was used as a source for the astrological predictions that appeared in every good almanac.*

The Mirror of Astrology

Since every man wants to know what may happen to him, I have really enjoyed the profession of Astrology, which depends on the celestial movements and from which many good things come, as Ptolemy has proved. However, I have resolved to have the

Nostradamus (1505–1566), undoubtedly one of the greatest astrologers, was also physician to Charles IX of France. Nostradamus became famous for his prophecies (*Centuries,* 1555).

natural Astrology printed for the satisfaction and contentment of every curious and learned man, showing accidental things, and all the Astrologers as well as the Philosophers agree that the Celestial bodies have governance and domination over earthly things; it is also quite true that man can, by his will, resist the inclination of the Stars and Planets, as Ptolemy writes. Aristotle proves that perilous points may be fled and avoided by means of prudence *(Sapiens dominabitur Astris)*. Every man should thus try to resist; it is written that Wisdom may be cured by the constellation and its evil effects if he sees himself inclined to such a fate; this is not a difficult thing in spite of the influence of the Stars over our birth. Free will is our own will, so that we can preserve ourselves from things that are contrary to it, as much by the grace of God, who is above everything, as by reason, so that nothing certain may be predetermined; whoever states the contrary deserves to be punished and his doctrine is useless, such as Necromancy, Geomancy, Edromancy, Aromancy, none of which have any foundation or value, and which are, for that reason, condemned by the Holy Church.

Moreover, I declare that the fixed stars, which are all known to the Astrologers, are 1122 in number; they are divided into 43 images or figures and 7 planets; that is why those who are born under these signs and influences are not obsessed by the respect of our free will.

The greatest secret of this Book
Is how to die and to
live well
And to live well and
die well
To try to achieve the most solid
good....

The Woman who is born in the month of December will be good natured, with a ruddy complexion; but the majority will be brunets, with black hair, fine eyebrows, blue, green, brown, or black eyes; some will have black hair, or any other color; she will have a good figure, will have a sign on her head or arm or hip; in her youth she will be neither black nor white, will be choleric, disbelieving, and will not confide in anyone; she will have power, will make enemies for herself, and will be so much inclined to anger that she may even slap someone or may wish to kill herself out of malice, which will in fact happen. She will go to great trouble to look after her body and will look at herself often in the mirror; she will keep her body in good order, her hair strong, and, being rather melancholy, she will sleep after meals; she will suffer from stomach and chest pains; she will suffer on account of her knees or have a disease in one of her limbs, will lose some teeth, and endure a serious illness at the age of twenty-three; she will be generally quarrelsome, will leave her country, will lose her paternal goods, and will have children; she will be prudent and full of good advice, will please everyone and have much pleasure, will love her good friends, will have many possessions at over forty years, and will live to seventy.

This work is written only out of curiosity and not to believe or determine certain things, since everything depends on the will of God who can do what he likes; nothing happens down here on Earth by chance alone, everything is led by Him who can do all things.

Nostradamus
The Mirror of Astrology

The Ship of Fools

From the time of its appearance in 1494, during the carnival season before Lent, The Ship of Fools, *written by the German humanist Sebastian Brant (1458–1521), has enjoyed phenomenal popularity both in Germany and in the rest of Europe. Numerous translations and plagiarisms testify to its success. The author causes all the fools of the land of Cockaigne to embark on a strange ship that is on its way to "Narragonien," the realm of folly.*

Representatives of every social class take their place on The Ship of Fools, *in which each chapter is devoted to a different fool personifying a particular human vice. Brant could not resist including a passage on the folly of wanting to know the sky.*

Of Attention to the Stars

Fool he who'd promise more than he
Can keep with full propriety,
More e'en than he'd desire to do.
Physicians well
may promise you
But many fools
will promise more
Than all the world
could hold in store.
For future things
one feels a bent,
What stars and
what the firmament
And what the planets'
course proclaims
And God's wise
providential aims.
They wish to know
and would discuss
Th' Almighty's plans
for all of us,
As though the stars
prescribed to you
What you should do
and should not do,
As though our Lord
were not much higher,
Not guided by
His own desire,
And letting even
Saturn-born
Be good and pure
as is the morn,

Sixteenth-century engraving of Sebastian Brant.

While Sun and
Jupiter have had
Some children
that are very bad.
A Christian true
should never heed
Base heathen arts
of any creed:
One can't by
scanning planets say
If this be our
propitious day
For business, war,
or marrying,
For friendship or
for anything.
Whate'er we've done,
where'er we've trod,
Our conduct should
rely on God.
He lacks a faith
in God's creation
Who trusts in
any constellation,
Believing certain
things are good
And so propitious
that we should
Start things of
moment only then,
And if they're not
accomplished when
The stars are lucky,
then delay,
Since that's an
unpropitious day.
And he who sports
no brand-new thing
At New Year's—
goes not forth to sing,
And hangs no greens
about the room,
He thinks he sees
approaching doom
As once Egyptians
seemed to fear.
And then, when comes
the bright New Year,
The one who gifts
does not receive
Will fear a year
when he must grieve
And other superstitious fancy,
Palm-reading,
e'en ornithomancy,
Or symbols, signs, and
books of dreams,
Or search for things
by Luna's beams,
Or black arts done with
pomp and show.
There's not a thing
men wouldn't know.
Who says:
"I've all that I desire!"
Lacks many things,
the brazen liar.
Not only do men prophecy,
From starry course
they also try
To read the smallest
things there are—
A fly's brain lit
by twinkling star,

The woodcuts illustrating *The Ship of Fools* (pages 165–7) contributed greatly to the work's success. Brant wanted even the illiterate to be able to learn from his satirical poem.

Whate'er'll be said
and what be done,
If you'll have luck,
if you'll have none,
Your health,
your very destination,
They'll read it
from a constellation.
With folly all
the world's insane,
It trusts the fool,
the most inane.
In calendars and prophecy—
In these the printers profit see,
They'll print
whatever one may bring,
Yes, e'en the
most disgraceful thing.
Such things go scot-free,
are believed,
The whole world
wants to be deceived.
If such art were observed and taught
And not with
dangerous evil fraught,
So that for man's soul
'twould be well,
As Moses did and Daniel,
Of wickedness
it would not savor,
'Twould have our
confidence and favor.
The cattle's death
they do divine,
The blight of corn,
the falling wine,
When rain will come,
and when 'twill snow,
When sun we'll have,
when wind or no.
The peasants like it—
good their reason:
It helps them plan
a prosperous season,
For they may hoard
their corn or wine
Till prices upward
may incline.
When Abraham read
books like these,
And sought Chaldean prophecies,
That consolation
then was gone
That God had
sent in Canaan.
Yet naught it is
but blasphemy
To treat such
matters frivolously,
To force th' Almighty we'd aspire
And bend His will to our desire.
God's love and favor
now have fled,
We seek the devil's art instead.
When God forsook the old king Saul
The devil heeded well his call.

Sebastian Brant
The Ship of Fools, 1944

French Folktales

Henri Pourrat (1887–1959) was a regional writer who devoted most of his life to collecting popular folktales and songs and classifying them by theme. If it had not been for Pourrat many of these tales probably would be lost. His French Folktales *is today a treasure for those interested in folklore.*

Sixteenth-century engraving by the German satirist Thomas Murner.

The Johnnies-in-the-Moon

There was once a stretch of countryside between mountain and plain. The reason I'm not telling you the name of the place is that I'd rather you didn't know it. One year there, in April, the frost brought by the "red moon" caused a disaster. It killed every bud on the vines.

The following Sunday, the wine growers met in their vat room to discuss the catastrophe. Everyone knows what *that* means, a get-together down in the cellar out there in wine-growing country—yes indeed, everyone know what *that* means....

That evening, somewhat worked up, they gathered on the square with the feeling that something just had to be done.

Now, according to the neighbors (who no doubt were simply jealous), these fellows weren't outstandingly bright and couldn't even have told you which month Our Lady's August feast day comes in.

Actually, it was just that they didn't have much luck. Any one of them falling on his backside would have been bound to break his nose.

You could tell they had bad luck because of all the misfortunes they suffered in this run-in of theirs with the moon. It all amounted to three or four good stories—and a lot more reliable ones, too, than the kind you get in a big-time politician's speech.

Anyway, these Johnny-in-the-Moon people had been fortunate enough to find a genuine brain in their very own town. Even as a child, this fellow had had ideas popping into his head the way pooperelles—excuse me, but that's what we call them around here—drop from a goat's behind. I'm talking about those sort of black lozenges that goats scatter ...when they go scrounging from bush to bush on a nice sunny day.

Ideas? That fellow had had them by the dozen, till at last the people made him their mayor. There among the mountain spurs and vineyards, he led all those good old folks along like a herd of donkeys.

So that Sunday, the whole commune was looking to their mayor to fix things up for them and make sure the frost would cause no more tragedies.

They all were huddled under the big elm, red-cheeked and hats askew, arguing, calling to each other, and generally carrying on.

"Silence!" shouted the mayor. "We've got to think rationally. Listen closely. It's the moon which caused the damage. If we manage to get rid of the moon, we'll save our harvest for all time, and long live the bottle!"

Now *that* was talking—*and* thinking. Every eye was on the mayor, every mouth hung open like a baby magpie's beak as they waited for what would come next.

"All right," the blacksmith still objected, "but how are we to get rid of the moon?"

"And what difference will it make?" cried someone else.

"Be quiet," said the constable. "Our mayor's got a good head on his shoulders. If something can be done, he'll find it."

"I don't say it won't cost us some work," the mayor continued. "Look: she's suspicious already. You see her up there, peeking at us and laughing to herself?"

The moon was just then poking her head out from behind the steeple, like a boy who's scrambled up to peer over a wall. Yes, she'd come to watch them through the branches of the elm. She wasn't quite round, but seemed instead to be leaning toward them, so bright, so white, and so peaceful....

"You see her? She's laughing at us, I tell you! But there must be a way to knock her down from up there. All we need's a little ingenuity. Ah, I'm already getting an idea. Perhaps we're not quite up to it this evening. But let's meet again at this time tomorrow. I want each and every one of you to bring his wine-making tubs!"

None of them could yet imagine what the mayor wanted with these tubs, to keep the moon from doing any more harm. But they'd bring them anyway. Their mayor was clever! Meanwhile, they all went for a good big glassful.

The next evening, they were all back with their tubs. And there were a lot, really a lot! The tubs filled the square, where usually you saw only a dandelion or two, a couple of rocks, or a few wisps of straw turning in the wind.

The moon came into view, seeming as before to be watching them and laughing. She was rounder than last evening, and above all, more threatening. All evening the air had been mild, smelling of flowers, warm earth, and new grass; but now it was growing chill. The clouds melted away and the moon shone up there in a clear, empty sky, dazzling and well able to finish blasting the vine buds. That fiendish moon was going to freeze everything up!

The mayor gave his orders. The constable—an old rascal as red as a cock and nimbler than a squirrel—was directed to set a first tub right in the middle of the square, then to place another one on top of it, and another on top of that, and so on and so on....

Up and up rose the tower of tubs. Quite carried away, the constable kept topping it off with yet another tub. Higher than the houses it was! Higher than the steeple!...

He was so high up there, good people, he was so high! Just like a weathercock on a church tower! They could hardly hear his voice any more—it was just a whisper, a distant sigh....

"One more! This time'll be it!"

But the tubs were all gone. They scrounged around everywhere, but not one was left. Every last tub had been brought from the cellars and pressed into service.

Up there in the sky the constable was hollering, "One more! One! Just one little tub! She's that close! I've got 'er, I tell you!"

In this 19th-century engraving peasants try in vain to capture the moon with their rakes. Such images are legion in popular folklore.

Everyone ran around like mad, searching in every corner, but there wasn't a tub to be found. It was enough to make you tear your hair out!

All at once the mayor had an idea. Ideas were his specialty, and he was hardly going to dry up in a crisis like that.

"We don't need the bottom one any more! Let's have that bottom one!"

They hopped to it.

What a clatter and crash! What shrieks! What cries! The din could have split the church walls. And right in the middle of it all...that old monkey of a constable came thundering down.

Still, once he hit the ground, he could not fall any lower. He got out of it with a few broken ribs and a sprained shoulder. Later on, though, those injuries never discouraged him from dropping in for a drink at one house after another. And he got off to a fine start that very evening, when they all whisked him right off for a good big glassful.

Another evening (this time he probably hadn't drunk quite *enough*) he was coming back from his rounds among the vineyards and their owners, stepping along bravely with his big red nose cleaving the wind. He was stumbling a bit over the rocks, and weaving somewhat from left to right, because it seemed to him that the path kept on suddenly straightening out in front of him. And what did he see when he got to the top of the hill?

Yonder, past the village, a moving, rising redness was glowing like coals when you open the oven door.

Heavens! He raced home, took down his drum, slung it around his neck, and grabbed the drumsticks. Off he charged through the streets, drumming the alarm.

"Fire! Fire! Fire!"

"Where? Where? Where?"

Everyone came out, shouting and slamming doors.

"By the pond! By the pond!" he answered, still battering away at his drum and heading that way on the double.

It was the moon—the moon rising as red as coals behind the black willows, and reflected in the water.

They got over all the excitement by all going for a good big glassful.

After that, they say, the constable had it in for the moon worse than ever....

Those two had a score to settle, he and the moon. He kept an eye on her. No, he wouldn't have her mocking him forever.

One evening he showed up at the mayor's place, panting for breath.

"Quick, Mister Mayor! She's touching the hill! Hang me if I don't get her this time!"

She did indeed look as though she was just resting on the hilltop.

They grabbed a wheat sack that was lying around and raced up there as fast as their legs would carry them.

But the hill was on the other side of the river and it was steep. When they finally got to the top, breathless and fainting, the moon had slipped away again.

They stalked the moon two or three times again like that, sack and all, but never caught her. To recover their spirits, they went for a good big glassful.

Now, the mayor was so clever that he at last realized one thing: "What with this game of ending up always Johnny-in-the-Moon, people around here no longer respect me as much as they used to."

He remembered the proverb, "You're better off calling the wolf a good friend."

"I'd be better off complimenting the moon," said the mayor to himself.

"We weren't very smart, were we, to try so hard to get rid of her! Who'd touch our seeds with moonlight?...

"Without the old moon to show us when to plant, our lettuces would all go to seed, and our radishes too, and our chervil. Who'd touch our woods with moonlight?...

"Wood not touched with moonlight, as the proverb has it, tends to get worm-eaten. And the moon shows when you're supposed to cut your nails or your hair. The moon! If there wasn't one, we'd have to invent her. Yes indeed, my friends, let's keep her, since we're lucky enough to have her!"

When that mayor felt like it, he could speak golden words.

This happened in front of the inn, one Sunday in May. The mayor spoke under the arbor, in the shade of many bottles, and his speech deserved to linger in every mind. To make sure it got all the way into their heads, each one who heard him downed another good big glassful.

One nice quiet evening a fortnight later, as the smith was returning from his vineyard by the path along the pond, he suddenly saw.... What do you think? None other than the moon that had fallen right there into the water.

Why, she had to be fished up and put back in the sky! They couldn't afford to lose her! The smith remembered the mayor's speech very clearly. Not having the moon any more would be a disaster!

So he had to think fast....

It didn't occur to him to look up and see whether the moon was still there. In that part of the country, ideas occurred only to the mayor. Anyway, he couldn't have thought of any such thing. There was no way the moon could be in the sky since he could see her distinctly, and very near, down in the pond among the willows.

So he ran to the village and brought everyone back—every Johnny-in-the-Moon thereabouts. What were they to do? How were they to fish her up?

"Somebody go for the village donkey," commanded the mayor. "Have the donkey drink the water. When he's drunk the whole pond, we'll have our moon

Woodcut of the sun and the moon.

and we'll hang her back in the sky."

They ran for the donkey and rushed him over as fast as they could. Being a good old donkey, he happily started drinking.

Just then a cloud passed by. Before you could say boo the moon was gone.

"Woe is us! The donkey swallowed her!"

They went down into the pond to drag the donkey out. But they tumbled in so roughly, with such shouts and carrying-on, that the donkey got frightened. He reared, kicked the water, dodged, and blinded them with spray. Then he got away....

They gave chase. All worked up, those Johnnies-in-the-Moon ran every which way after him, till at last their mad pursuit ran him to earth in the square with the big elm. There, right in front of the inn, they jumped on him.

They were so excited, you know, they didn't even hesitate. They stunned the donkey with a whack on the head and opened up his belly.

There was no moon in there.

"Well, by thunder, that galloping donkey scattered his droppings all over, and he dropped the moon somewhere too. Who's to know where? We'll probably never see her again!"

Everyone went back home, downcast.

But as the smith came to the apse of the church, he let out a great cry.

"Here she is! We've found her! She's in here!"

And there she was, indeed, in an old cistern squeezed in between the round wall and a few hovels. The cloud had sailed by and she'd just come out again when he'd come on the cistern and spotted her.

"The donkey left her there! We've got to get her out right away!"

Everyone ran up again behind the mayor. The policeman brought the donkey's rope. It was pretty frayed, but long enough. They tied a hook to it and tossed the hook into the cistern.

A dozen of them began pulling.

The hook, meanwhile, had caught on a stone at the bottom. It was completely stuck and wouldn't move.

More men joined in, till there were twenty or so heaving away hard enough to skin their hands.

Suddenly the rope broke and they tumbled over together on their backs. ... There above them rode the moon, right back in her place among the clouds.

"Well, there she is, after all that work! We sent her back where she belongs!"

That was the end of their tangle with the moon. Those Johnnies-in-the-Moon were so pleased and proud that evening, they got drunker than thrushes in grape-harvest time. After all, they could hardly *not* go and celebrate with a good big glassful.

Henri Pourrat
French Folktales, 1989

PELLERIN & Cie, imp.-édit.

RÊVE ÉTOILÉ

IMAGERIE D'ÉPINAL, N° 629

A coup sûr, Athanase Fromageot se croyait supérieur à la modeste et positive profession dont vous voyez près de lui un des principaux attributs.

A force de regarder le ciel en le prenant à témoin de la grandeur de sa résignation, il contracta un violent rhume de cerveau.

Pour lequel une bonne femme, sa cliente, lui conseilla une infusion dont elle lui promit merveille.

Athanase se fit servir le remède un peu avant de se mettre au lit, comme il lui avait été recommandé pour qu'il eût toute son efficacité.

Puis s'étant couché, il fit un dernier éternûment et s'endormit d'un lourd sommeil.

Tout-à-coup une lueur intense vint l'éblouir. C'était une Comète qui lui passait délicatement sa queue sur la figure à l'effet de le prévenir de la visite de la Lune et du Soleil qui se précipitaient vers son lit à qui arriverait premier.

Un choc eut lieu où le Soleil perdit quelques rayons et où la Lune eut le nez écrasé. Ces astres venaient inviter Athanase à quitter le monde inférieur pour des régions plus dignes de sa haute intelligence. Quand ils prirent congé, leurs dégâts étaient réparés comme par enchantement.

Cette invitation répondait si bien aux idées habituelles d'Athanase que son parti fut vite arrêté et qu'il quitta aussitôt sa boutique, sans prendre seulement la peine d'en fermer la porte.

Et comme par miracle, il se trouva soudain à califourchon sur un soufflet ailé qui, crachant et soufflant, le transportait avec une merveilleuse rapidité vers les régions supérieures.

Mais l'enthousiasme d'Athanase ne tarda pas à se calmer quand il vit que le ciel s'obscurcissait et qu'il avait à traverser d'épaisses couches de brumes glacées. Il se félicita d'avoir songé à se munir d'un parapluie.

Le soufflet se dirigeait vers la Lune qui, au départ, était dans son plein. Mais celle-ci s'étant pendant son voyage réduite en croissant, le soufflet la manqua. La déception ne fut pas partagée, car la Lune eut un sourire gouailleur.

Et notre pauvre voyageur en fut réduit à aborder sur une planète inconnue dont l'aspect froid et stérile ne lui présageait rien de bon. Il commençait à croire à une mauvaise plaisanterie.

Et il eut une grosse émotion en voyant tout-à-coup surgir devant lui une espèce de gendarme qui lui intima l'ordre de se rendre immédiatement au bureau des départs, l'accès du pays étant interdit aux étrangers. Il devait avoir disparu dans une heure sous peine des plus cruels supplices.

Athanase troublé, demanda à un naturel ce que c'était que ce bureau des départs.

— C'est, lui fut-il répondu, cette maison que vous voyez là-bas. A la porte, vous sonnerez. Une trappe s'ouvrira sous vos pas et vous irez où personne ne sait. Pas drôle.... mais vous n'avez pas le choix.

C'était vrai, il n'avait pas le choix! et mieux valait encore l'inconnu que les supplices promis en cas de désobéissance. Seulement on ne peut pas dire que, arrivé à la porte, la sonnette en main, Athanase n'hésita pas. Enfin il tira le cordon...

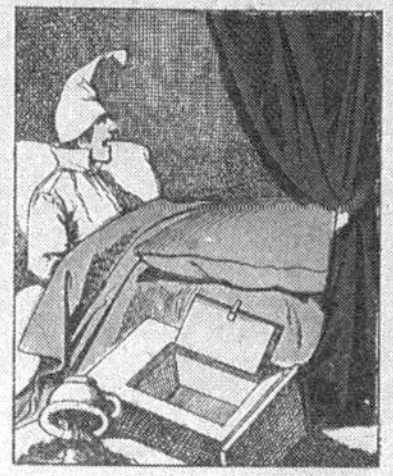

Un fracas épouvantable se produisit et Athanase se réveilla dans son lit; tout simplement, dans son émoi, il venait de bousculer sa table de nuit: le remède de la bonne femme avait bien agi, car notre incompris se trouvait guéri à la fois de son rhume de cerveau et de ses folles chimères.

PELLERIN & Cie, imp.-édit.

LE ROI DE LA LUNE

IMAGERIE D'ÉPINAL, N° 932

Georges a fait l'école buissonnière : il est si fatigué qu'il s'endort dans un champ de blé.

Il voit alors s'approcher de lui un homme singulier qui a de grandes ailes attachées aux épaules.

Cet homme lui dit : « je suis le Roi de la Lune et je vais t'emmener là-haut avec moi. »

Georges ne demande pas mieux; il s'installe sur les épaules du Roi de la Lune, qui s'envole aussitôt.

Arrivés dans la Lune, les voyageurs sont salués par une foule d'enfants qui marchent à quatre pattes.

Qu'est-celà? demande Georges. — Ce sont, répond le Roi, les petits gourmands de la Terre; je les condamne ici à ne manger que de l'herbe.

Plus loin, on rencontre d'autres enfants et Georges s'aperçoit qu'ils ouvrent la bouche sans parler parce-qu'ils n'ont point de langue.

— Ceux-là, dit le Roi, sont les menteurs et les bavards; j'ai toutes leurs langues dans mes poches : c'est ainsi que je les punis !

A quelques pas, on voit d'autres petits condamnés qui sont enfermés jusqu'au cou dans de grands sacs.

Le Roi de la Lune dit à Georges : « Voici les batailleurs qui frappent sans cesse leurs camarades ou tourmentent les animaux ».

Georges se doute de quelque chose; il essaie de se sauver, mais le Roi de la Lune le retient par sa blouse.

On arrive dans une allée d'arbres, et à chaque arbre est attaché un petit garçon ou une petite fille.

— Voilà ta place, dit le Roi; ce sont tous les paresseux qui font l'école buissonnière au lieu d'aller travailler à l'école.

Georges voit un nègre administrer une volée de coups de verges à tous ces enfants : quand il a terminé, il le saisit lui-même.

Georges pousse un grand cri et se réveille dans le champ de blé où il s'est endormi, heureux d'en être quitte pour un rêve.

Mais il se dit : le Roi de la Lune a raison de punir les paresseux et ce m'est une leçon ! Sur cette réflexion, il s'empresse de se rendre à l'école.

The cartoons on the preceding pages illustrate 19th-century moral tales for children. In the translations below, each slash indicates a new illustration.

Starry Dream

Athanase felt far superior to his trade. / One day he developed a head cold. / One of his clients suggested a herbal tea that could work wonders. / He took it just before going to bed. / He sneezed and fell asleep. / Suddenly, he was blinded by a comet, which let its tail gently stroke his face to warn him that the sun and the moon were racing to reach him. / As they clashed the sun lost some of its rays and the moon's nose was flattened. They had come to invite Athanase to leave the lower world and to attain heights more suited to his great intelligence. / Athanase found the idea so appealing that he left his shop without even locking the door. / Magically and swiftly he was transported to higher regions astride a pair of bellows. / He was less enthusiastic when he saw that he would have to go through thick layers of freezing mist. Luckily, he had brought an umbrella. / Athanase approached the moon, which was full, but missed it because it shrank to a crescent. The moon smiled cheekily. / Our poor traveler was forced to land on an unknown planet that was cold and sterile. / He was surprised to see a policeman arrive and tell him that the planet was off limits to foreigners and he would have to go at once to the departure office. / Athanase was worried and asked one of the natives about the departure office. He was told it was a house in which a trapdoor would open beneath him and he would fall to some unknown place. / The alternative was some horrible form of torture, so he went up to the house with trepidation and pulled the cord. / There was a tremendous noise, and Athanase woke up in his own bed. The woman's remedy had worked: He was cured both of his head cold and his foolish dreams.

The King of the Moon

While playing hooky, George falls asleep in a cornfield. / A stranger approaches him. / The stranger says: "I am the King of the Moon and I shall take you there with me." / George holds on to the King's shoulders and they fly off. / On the moon they are greeted by children who are walking on all fours. / "Who are they?" asks George. The King replies, "They are the young greedy children of the Earth. Here I condemn them to eat only grass." / They encounter more children, who open their mouths without speaking because they have no tongues. / "Those are the liars and the gossips," says the King. "I have all their tongues in my pockets." / They see other prisoners who have bags tied around them up to their necks. / The King of the Moon says to George, "These are the fighters who always hit their friends or torture animals." / George tries to escape but the King holds on to him. / They arrive at a row of trees to which children are tied. / "Here," says the King, "are all the lazy children who play hooky instead of going to school." / George sees a man whip the children with a rod and then feels himself being grabbed. / He shouts out loudly and wakes up happy that it was only a dream. / He says to himself, "The King of the Moon is right to punish the lazy ones and I've learned my lesson," and he hurries back to school.

Engraving of the month of June, from *The Great Calendar and Compost of the Shepherds.*

Popular Meteorology

Although modern astronomical science has little to learn from the popular beliefs of yesterday, the same is not true for meteorological science. Comparison of selected empirical observations of the past with the models that meteorologists are now developing on computers allows us not only to begin to compile a history of climates but also to attempt to ascertain what universal truths may exist in the accumulated observations that are expressed in the form of popular sayings.

The Feast of Saint John the Baptist

[John the Baptist] is the saint of the summer solstice, whereas the other John, the Evangelist, is the saint of the winter solstice. He represents and sanctifies the increasing light of summer while the other John represents the declining light of winter. This gives rise to a saying:

John and John
divide the year.

The identification of the Christian Feast of Saint John with the pagan solstice was familiar to our ancestors:

The night of Saint John's Day
is the shortest one of the year.

Or, on the contrary:

Around Saint John's Day
we get the longest days.

It is not surprising to see this day linked to superstitious practices or at least to customs that are pagan in origin. Take that of the fires, for example:

On Saint John's Day
the fires are great.

This refers of course to the fires of the sun, but also to those that are lit on the hilltops and around which the young people dance....

The witches and warlocks collected medicinal plants on the night preceding the celebration of Feast of Saint John, the night of the 23 to 24 June:

The herbs of Saint John's Day
retain their power all year long.

Even stranger is the superstition evoked by another dictum:

If you want a good harvest you must
sleep on your manure on the night
before Saint John's Day.

It was sorcery in its pure form, if one may say that, that manifested itself in this saying:

If Saint John finds a sitting hen
either animals or people will die.

Worse still, in one variant Saint John "passes by and wrings the neck" of the hens he finds sitting on their eggs.

One may discern in these somber threats the survival of some pagan mystery....

25 December: Christmas Day

The custom of celebrating the birth of the Infant Jesus is a relatively recent one. It did not appear before the 4th century and was not widely observed until the 7th or 8th century.

The commemoration of this birth on 25 December has no historical value: We do not know the true date of Christ's birth. But it does have a very strong symbolic value: The 25th of December, three days after the winter solstice, which marks the darkest point of the astronomical year, is the first day on which the year is "reborn" from the night, moving toward light.

The French word for Christmas, Noël, comes originally from the Latin, *natalis dies* (day of birth), which becomes *natalis* by abbreviation, then *Nadal* or *Nau* in Provençal, and *Nouèl, Noé,* or *Noël* in Old French. The latter form is the only one that has survived to the present.

The word is normally masculine. Etymologically, it refers to the day of Christ's birth and not to the feast celebrated on that occasion.

Christmas falls on a fixed date (25 December), but obviously not on the same day of the week.

More than any other day, Christmas Day is eminently a "predictive" day. The minute observation of the weather on that day enables one (or is said to enable one) to say what the weather will be like until the following Christmas.

The days between Christmas
and the Kings
show the weather of the twelve months.
Twelve days from Christmas
to the Kings,
the weather of the twelve months.
Watch the appearance
of the twelve days of Christmas,
for according to these twelve days,
the twelve months will have
their course.

These twelve days are (or used to be) called the "male days" or the "days of fate," depending on the region. Each of them symbolically "governs" the month that occupies the same place as it does in the miniyear that follows Christmas. We see here the system of cosmic "correspondences" according to which the small (or microcosm) contains the large (or macrocosm).

It is especially the confrontation between the two most sacred days of Christianity, Christmas and Easter, that plays a prophetic role:

Christmas on the balcony
means Easter by the fireside.

Or, in two variants:

Christmas by the bush or on the steps
means Easter by the fireside.

Without exhausting the available material, one may cite a few more examples:

Midges at Christmas,
icicles at Easter;
Icicles at Christmas,
midges at Easter.
He that warms himself
by the sun at Christmas
burns the Christmas log at Easter.
He that seeks the shade at Christmas
seeks the hearth at Easter.

The shady spot (*ombrier*) in Provence, and in southern France, is the corner of the garden that is always shielded from the sun; the *foger* (hearth) is an old southern French form of the word *foyer*....

From Brittany comes a very strange saying:

At Christmas "les limas,"
at Easter "les grouas."

How are we to interpret these *limas?* As snails [*imaçons*]? It may also be a reference to the Old French word, which became slang, where *limas* were light linen shirts. But what could the *grouas* be? That is a mystery...

Another saying evokes the two culinary specialties of these festivals: the Christmas cake and Easter eggs:

When one eats the cake in the heat
one eats the eggs by the stove.

Even more amusing is another version of the same saying:

Illuminated manuscript page for December (above) from *The Hours of the Duchess of Burgundy.* Right: Lithograph illustrating a snowball fight on Christmas night.

When one eats bouquettes at the door,
one eats cocognes by the fireside.

Cocognes evidently are Easter eggs, and *bouquettes* are Christmas rolls.

Christmas Day, but above all the night of Christmas Eve (the night of 24–25 December), are important indicators of future harvests. A dry cold and snow are favorable signs.

At Christmas, a hard frost
heralds the fine ears of corn.
Christmas of frost, grainy crop.
Soft Christmas, soft crop.
If it freezes on Christmas morning
the floor will be full of grain.
If there is frost at Christmas,
there will be ample cider.

The phase of the moon (full or new) during which Christmas falls, and, more generally, the lunar cycle that begins at Christmas, often appear in sayings that can be contradictory.

When Christmas arrives "without a moon," that is, at the new moon, it is a bad sign:

Christmas without a moon,
of two cows, one will eat one.
Christmas without a moon,
of a hundred lambs, only one
will remain.
Christmas without a moon,
three sheaves will not be worth
a single one.

We should understand that three sheaves from the coming harvest (if there was no moon at Christmas) will not be as full of grain as even a single sheaf in a good year. The only consolation is that:

Christmas without a moon
means a year of plums.

It is scarcely less unfortunate for Christmas to fall at a full moon:

Bright night of Christmas,
clear swath.

That is to say, little corn. Similarly:

When Christmas is starry,
there'll be much straw and little corn.
Moonlight at Christmas
means empty fields the following year.
When Christmas is accompanied with
brightness,
sell your oxen, and buy corn.

That is because there will be a shortage of corn the following winter. This series could be summed up briefly by the following formula:

Bright Christmas, somber prospects.

Apart, that is, from a few fruit:

The brighter the moon is at Christmas,
the more fine apples there'll be.

The wind on Christmas Eve is also taken into account:

A wind that blows at the end
of midnight mass
will dominate the following year.
Strong wind at Christmas,
abundant fruit.

In any case, to come back to everyday reality, there has to be a Christmas and a winter:

Christmas carries winter in a bag,
either in front or behind.
Between Christmas and Candlemas,
there are no more ploughmen.
He who has salted his pork
will not go hungry at Christmas.

This is because Christmas, the festival of plenty for the rich can also be a very cruel day for the poor!

Jacques Cellard and Gilbert Dubois
Sayings for Fine and Rainy Days

Like peasants, sailors have long suffered the hazards of the changing climate. Through the years they developed a meteorological lore that is expressed in popular sayings.

Proverbs and Sayings of the Ancient Mariners

The Ancients knew about such matters, having long gazed at the sky and the sea. The toilers of the sea knew about these things, repeating day and night the accumulated lessons learned on all the seas of the globe. They condensed the precious reserve of seafaring experience into proverbs and sayings that were passed on to young people.

In their time there were neither weather forecasts nor radio. Ancient mariners had to set sail whatever the conditions. They knew how to predict good or bad weather and gales, by means of the moon, the sun, the clouds, and the stars, lightning, stormclouds, and mists.

Thanks to these proverbs and sayings many sailors even today can predict what the weather forecast cannot tell them.

Predictions of Good Weather

Balls of clouds mean winds from above.

A light arch of cloud, ten fingers wide to the south is a clear sign of good weather.

Fog in the valley,
fisherman, bring in your nets.

When the green down appears
the north wind will blow.

If the sea is ruffled by the wind
there will be upsurges of wind.

If the lower part of the crescent
moon appears for four days,
there will be several days of fine weather
and the topmast will not break.

Large sun on waking: light wind;
small sun on waking: strong wind.

Brilliant clear moon
in its first crescent
or at the full moon:
good watches for the sailor.

White frost at the crescent moon,
fine weather.

White frost at the waning moon,
rain for three days.

Northwest wind, broom from the sky,
fine weather after a rainbow.

Predictions of Bad Weather or of Strong Winds

Sun like the moon,
wind from above or mist.

Sun encircled in twenty-four hours,
time to bring in the sail.

Sun in a shroud,
sailor, prepare your long coat.

Dappled sky and mares' tails
will make even the largest vessels
tie down their sails.

Cat's whiskers in the clouds
announce stormy winds.

Northwest winds that get us wet
are not worth a hood.

The gentle southwester:
When he gets angry, he becomes crazy.

An encircled moon
never brings down a topmast

Drawing by Marc Berthier of a ship in a storm (left). The engraving above is an example of a sailor's ex-voto.

since the captain who sees it,
waits for good weather.

Round lower point of
the crescent moon,
bad weather on land and on sea.

Circle round the moon in the evening,
wind and rain at midnight
you will feel and see.

Yellow peelike moon,
the seas will be troubled.

A sky covered with fleecy clouds,
like a woman's makeup,
does not last very long.

Joe Klipffel
Predicting the Weather with the Help of Sailors' Sayings

The Excesses of the Weather

The condition of the weather is integral to daily life. Weather is subject to many extremes, and when it becomes unsettled people can be overcome by panic. Even today we are often powerless in the face of the elements and must submit to the sometimes disastrous whims of the sky.

Engraving of a train blocked by snow, January 1912.

A rare sight: Niagara Falls frozen solid.

The cold spell of January 1912 was particularly harsh: A transatlantic liner arrived in New York covered with ice.

In the engraving above, a shepherd watches helplessly as his flock is struck by lightning.

Nineteenth-century engraving showing a farm machine struck by lightning.

The Festivals of Saint John's Day

It is a commonly held view that festivals held on the Feast of Saint John are to a large extent remnants of ancient solar cults. Yet if one looks closely at this notion one sees few, if any, traces of those archaic cults. The history of these Protestant festivals is more subtle than it might seem.

Summer Solstice, Winter Solstice

The solstices are the two moments of the year at which the sun is farthest from Earth's equator. At the summer solstice, in any given place, the sun rises higher in the sky at noon than on any other day of the year, and at the winter solstice it passes lowest of all at the meridian. In the northern hemisphere the solstice occurring between 21 and 23 June, depending on the year, marks the beginning of the summer and the longest day; the other, between 22 and 24 December, marks the start of winter and the shortest day. The correspondence is reversed in the case of the southern hemisphere. One may ask, What is the origin of this term, the solstice? It comes from the Latin *solstitium*, made up of *sol*, the sun, and *sistere*, to stop. But why is there this idea of stopping? This is because at the moment at which it reaches its highest or its lowest position in the sky, the sun seems to be fixed there; around the longest and the shortest days the variation in the length of the day is very small. That is why ancient scholars had so much difficulty in determining the precise moment of the solstice and why celebrating the winter solstice on 25 December is not a serious error.

The Festivals of Saint John

Christianity does not seem to have retained many manifestations of the solar cults. Yet there are many solar images of Christ, in particular, assimilations of Christ with the rising sun. Saint Luke [1:78–9] recalls that Saint John the Baptist's father, Zacharia, foretold the coming of Jesus: "God, who will send down to us the visit of the rising sun, in order to

The fires of the Feast of Saint John are a living tradition. Men dance around a fire in an early-20th-century photograph (left). Above: A bonfire in celebration of the summer solstice, 1934.

illuminate those who are in the darkness and the shadow of death"; the first Christians saw a prophetic image of Christ in the "sun of justice," which was announced in the prophetic book of Malachi. But it was in the liturgy and in religious service that the traces of the influence of the solar mythology may best be discerned. The first Christians turned to the East while praying, and the first churches, like the Greek and Roman temples, were built with their facades facing the East. This caused difficulties for the faithful coming to church and for the celebrant. From the beginning of the 4th century, it was no longer the facades but the apses that faced East. According to the earliest baptismal rites, the catechumen began by renouncing the Devil; facing West, the direction of the approach of night, and then turning eastward, from where

In the engraving above the inhabitants of a Brittany village on Saint John's Day are seen walking around the burned-out bonfire, throwing a pebble onto the embers while reciting a prayer. In an earlier engraving (right) shepherds and shepherdesses dance around the fire.

the day appears, he accepted Christ.

If Christianity was able to incorporate and transform the pagan festivals that marked the winter solstice, the moment at which the sun begins to rise again in the sky, having reached its lowest point in the sky, and to replace them by the celebration of Christ's birth, it was less successful with the summer solstices. Many of the folklore traditions of the festivals celebrating Saint John's Day on 24 June have pagan and mythical overtones.

On 23 June at sunset a pyramid of firewood is constructed on the village square. The priest arrives in a procession to set it alight and the heads of the families pass through the fire the bunch of flowers that will be nailed to their stable doors the next day before dawn. Then the young people of the village could dance around the fire and also jump over the embers, the remains of which would later be brought home. That same evening, the men drag to the top of a hill an enormous bale of straw, pierced by a long guiding rod, set it alight, and accompany it as it rolls down. When the wheel of fire is halfway down the hill, the women and girls who are waiting there greet their men and the fire with cheers.

In the mountains of France there is the custom of climbing to the top of hills before dawn to watch the sun rise. When the sun appears shouts of joy go up and are repeated in the distance; in the valley the churchbells ring and everyone gets up. Those who were sent up to watch for the sun come back to the village carrying bunches of aromatic herbs, which they distribute for their healing properties.

It is obvious that it is the sun being celebrated throughout Europe on this occasion. However, even though there was at least one pagan festival dedicated to the winter solstice, the one that the Roman emperor Aurelian in the 3rd century AD decided to consecrate to the "invisible sun" and which the Church transformed into Christmas, there is no known festival of the summer solstice. The origin of these traditions may have to be sought in the rites of rural magic and the sun may only play a secondary role in them, since it guarantees an abundant harvest. Coming as they do from further back in time, and benefiting from peasant roots that are deeper and more universal than the celebrations of the winter solstice, one can see why the festivals of summer resisted attempts at recuperating them more successfully. Just as it turned 25 December into the celebration of the birthday of Jesus, Christianity turned 24 June into the feast of the birth of Saint John, not the Evangelist but Saint John the Baptist, who baptized Jesus. A legend said that Saint John the Baptist's birth preceded in the year that of Christ, as if the message of one had announced the message of the other. Placed at the times of the two solstices in this way, the celebration of the two births divides the year into equal halves, with that of Saint John preceding that of Jesus. The amusing side of this is that in the arguments put forward to justify the date of 24 June, the Church employed a solar image. Speaking of Jesus, whose humble precursor he said he was, Saint John the Baptist is supposed to have said: "He must increase and I must decrease." In fact, after the summer solstice the days do grow shorter.

Wherever one is on Earth the sun rises to various heights in the sky, but this phenomenon does not have the same influence on the length of the day. At the equator the days and nights are the same length all year round. Both are always twelve hours long. At the poles, on the other hand, a day of six months follows a night of six months. It is at median latitudes, those of the temperate zones, at which the variation in the height of the sun is accompanied by a sufficient and reasonable variation in the length of the day, that the solstices have struck the popular imagination most deeply.

Jean-Pierre Verdet

Further Reading

Aristotle, *On the Heavens*, Loeb Classical Library, Harvard University Press, Cambridge, 1939

Brant, Sebastian, *The Ship of Fools*, Columbia University Press, New York, 1944

Cassirer, Ernst, *The Individual and the Cosmos in Renaissance Philosophy*, Barnes and Noble, New York, 1963

Eliade, Mircea, *A History of Religious Ideas*, 3 vols., University of Chicago Press, Chicago, 1981–5

———, *Myths, Dreams, and Mysteries: The Encounter Between Contemporary Faiths and Archaic Realities*, Harper and Row, New York, 1979

———, *Studies in Religious Symbolism*, Princeton University Press, 1991

Flammarion, Camille, *Popular Astronomy*, Chatto and Windus, London, 1894

Herodotus, *The History*, University of Chicago Press, 1987

Hesiod, *Theogony*, Oxford University Press, New York, 1966

Ions, Veronica, *Egyptian Mythology*, Peter Bedrick Books, New York, 1983

———, *Indian Mythology*, Peter Bedrick Books, New York, 1984

Kenton, Warren, *Astrology: The Celestial Mirror*, Thames and Hudson, London, 1974

The Koran, N. J. Dawood, ed., Penguin Classics, New York, 1956

Lévi-Strauss, Claude, *The Raw and the Cooked: Introduction to a Science of Mythology*, vol. 1, University of Chicago Press, 1983

Livy, *The Early History of Rome*, Penguin Classics, New York, 1960

Michell, John, *A Little History of Astro-Archaeology*, Thames and Hudson, London, 1989

The New Testament, New Revised Standard Version, Thomas Nelson Publishers, Nashville, 1989

Parrinder, Geoffrey, *African Mythology*, Peter Bedrick Books, New York, 1986

Pliny the Elder, *Natural History*, 10 vols., Loeb Classical Library, Harvard University Press, Cambridge, 1938–62

Pourrat, Henri, *French Folktales*, Pantheon Books, New York, 1989

Ptolemy, *Tetrabiblos*, Loeb Classical Library, Harvard University Press, Cambridge, 1940

Seneca the Younger, *Naturales Quaestiones*, 2 vols., Loeb Classical Library, Harvard University Press, Cambridge, 1971–2

Sesti, Giuseppe Maria, *The Glorious Constellations: History and Mythology*, Abrams, New York, 1991

Tacitus, Cornelius, *Agricola and the Germania*, Penguin Classics, New York, 1971

The Vedanta Sutras, G. Thibaut, ed., Gordon Press, New York, 1974

List of Illustrations

*Key: **a**=above, **b**=below, **l**=left, **r**=right*

Abbreviations: BN=Bibliothèque Nationale, Paris; MATP=Musée National des Arts et Traditions Populaires, Paris; MC=Musée Carnavalet, Paris

Front cover Amédée Guillemin. Shower of shooting stars on 27 November 1872. Lithograph in *The Sky*, 1877
Spine The Sun. Engraving in Michel Maier, *Atalante Fugiens*. Private collection
Back cover Representation of a lunar eclipse in Apianus, *Astronomicum Caesareum*, 1540. Library of the Paris Observatory
1–7 Engravings in Stanislaw Lubieniecki, *Theatrum Cometicum*, 1667, vol. 2. BN
9 Pissing on the moon. Painting by Brueghel, 16th century. Van der Berg Museum, the Netherlands
10 Observation of the sky. Colored engraving in Johannes Hevelius, *Selenografia sive Lunae Descriptio*, 1647. BN
11 Sun between two moons, seen in 1157. Engraving in Conrad Lycosthenes, *Book of Wonders*, c. 1550. Basel, Switzerland
12–3 Richard Tongue, *Stonehenge*. Painting, 1837
13a Prehistoric stela from Rocher-des-Doms, France. Musée Calvet, Avignon
14a Course of the moon in relation to the sun. Engraving in Apianus, *Cosmographia sive Descriptio Universi Orbis*, 1524
14b Table of the movements of the sun. Engraving in *Breviary of Love*, Provençal codex, 13th century. Library of the Escorial, Madrid
15 Table of the movements of the moon in relation to the sun. *Ibid.*
16 Signs seen in the sky on 18 March 1716. 18th-

141 Appearance of a comet on 12 November 1577. 16th-century woodcut
143 Engraving of an eclipse in Sacrobosco, *Treatise of the Spheres,* Paris, 1535
144–5 Watching the eclipse of 17 April 1912. Photographs, Paris
146 Anonymous. Meteor in Italy in 1937. In *La Domenica del Corriere*
149 Gustave Doré. Saint Elmo's fire. Engraving, 19th century
150 Saint John writing the Apocalypse, as dictated to him by the Virgin. Engraving
152 Albrecht Dürer. *The Apocalypse: The Signs of the Judgment.* Woodcut, late 15th century. BN
154 *Brumaire,* from the revolutionary calendar. Engraving. MC
155a Anonymous. Calendar for the year II, 18th century. MC
155l Anonymous. Calendar for the year II, 18th century. MC
156–7 Anonymous. Revolutionary calendar, 18th century. MC
159 *The Print Seller.* Engraving after a drawing by Jules David, 19th century
160 Woodcut from the *Shepherd's Calender,* 16th century
161 Woodcut from the *Almanac of the Good Farmer.* Troyes, France, 1792
162 Anonymous. Portrait of Nostradamus. Engraving
164 Portrait of Sebastian Brant. Engraving, 16th century
165–7 Woodcuts in Sebastian Brant, *The Ship of Fools.* Basel, Switzerland, 16th century
168 Engraving in Thomas Murner, *Logica Memorativa.* Brussels, 16th century
170–1 Men raking the moon. Engraving, 19th century
173 The sun and the moon. Woodcut
174 *Starry Dream.* Print, c. 1860, Epinal, France. Private collection
175 *The King of the Moon.* Print, c. 1860, Epinal, France. Private collection
177 The month of June. Engraving in *The Great Calendar and Compost of the Shepherds*
180 December. Miniature in *The Hours of the Duchess of Burgundy.* Musée Condé, Chantilly, France
181 Anonymous. Snowball fight on Christmas night. Lithograph in *The Illustrated London News,* 1860
182 Marc Berthier. Ship in a storm. Drawing, 19th century
183 Sailor's ex-voto. Colored engraving, early 19th century
184a Train stopped by the snow. Engraving in *Le Petit Journal,* 28 January 1912
184b Niagara Falls frozen over. Photograph
185 Cold spell in America. Engraving in *Le Petit Journal,* 28 January 1912
186 A flock of sheep struck by lightning. Engraving in the illustrated supplement of *Le Petit Journal*
187 A farm machine struck by lightning. Engraving, 19th century
188 Men dancing around the fire. Photograph, early 20th century
189 Summer solstice bonfire. Photograph, 1934
190 Saint John's Day bonfire in Brittany, France. Engraving, 19th century
191 Shepherds and shepherdesses dancing the farandole. Engraving

Index

Page numbers in italics refer to captions and/or illustrations

Acknowledgments

For their help in producing this work, the publishers wish to thank Les Editions Belin, Presses Universitaires de France, the Musée National des Arts et Traditions Populaires, the Société Française d'Astronomie. Dr. Steven L. Beyer provided technical assistance in the translation of the sky maps into English

Photograph Credits

All Rights Reserved 51, 96, 97, 141, 149, 152, 164–7, 177, 182, 195, 198, 200. Archiv für Kunst und Geschichte, Berlin 45, 46, 47, 56, 57, 68, 120–1b. Archives Gallimard, Paris 11, 26, 34a, 34b, 35, 44, 48–9, 72, 74l, 75r, 84a, 84b, 86–7, 88, 89, 94, 109, 114–5, 116–7, 118–9, 124, 126, front cover. Artephot, Paris 39. Artephot / Nimatallah, Paris 25. Artephot Percheron 91. Bibliothèque du Musée de l'Homme, Paris 52, 58. Bibliothèque Nationale, Paris 10, 14a, 24, 65a, 65b, 78a, 78–9b, 80, 81, 82, 83, 85, 92, 105, 125, 127, 140. Bridgeman Art Library, London 22. Bulloz, Paris 67, 76, 77, 150, 155a, 155b, 156, 162. Charmet, Paris 19r, 21, 29, 42a, 42b, 50, 59, 61, 87, 95, 104, 144, 146, 174, 175, 181, 183, 184a. Dagli-Orti, Paris 14b, 15, 18, 19l, 20, 27, 30–1, 32, 33, 36l, 36r, 37, 53, 69, 70b, 74–5, 90. Edimedia, Paris 40–1, 139, 154. Explorer Archives, Paris 43, 101a, 101b, 102, 143, 168, 171, 173. Fotomas Index, London 12–3. Giraudon, Paris 62–3, 66, 108, 113. Harlingue-Viollet, Paris 191, 193. Lauros-Giraudon, Paris 180. Marc Garanger, Paris 111. Musée Calvet, Avignon 13. Musée National des Arts et Traditions Populaires, Paris 28, 93r, 159, 161. Musée des Beaux-Arts, Brussels 96. Musée des Arts Océaniens et Africains, Paris 123. N.D.-Viollet, Paris 192. Photoresources, Canterbury 70a, 71, 73, 93b. Private Collection 16. Réunion des Musées Nationaux, Paris 54, 55a, 55b, 103, 112. Roland and Sabrina Michaud, Paris 121. Scala, Florence 17, 99, 107. Sirot / Angel, Paris 184b. Tapabor, Paris 185, 186, 187a, 187b, 188, 189. Viollet, Paris 38, 144, 145, 160, 190

Text Credits

Grateful acknowledgment is made for use of material from the following: Brant, Sebastian, *The Ship of Fools*, Columbia University Press, New York, copyright 1944 by Sebastian Brant (pp. 164–7). From *French Folktales* by Henri Pourrat, translated by Royall Tyler. Translation copyright © 1989 by Royall Tyler. Reprinted by permission of Pantheon Books, a division of Random House, Inc. (pp. 140–2, 168–73). Scripture quotations from *The New Revised Standard Version of the Bible*, copyright 1989, by the Division of Christian Education of the National Council of the Churches of Christ in the U.S.A., and used by permission (pp. 150–3)

Jean-Pierre Verdet is an astronomer at the Paris Observatory, where he has studied the physics of the solar corona and the atmospheres of Jupiter's moons as seen through the infrared spectrum. For the past fifteen years he has concentrated on the history of ancient astronomy, and he formed a multidisciplinary team to translate and publish astronomical texts of antiquity and the Renaissance.

Translated from the French by Anthony Zielonka

Project Manager: Sharon AvRutick
Editor: Harriet Whelchel
Typographic Designer: Elissa Ichiyasu
Editorial Assistant: Jennifer Stockman
Design Assistant: Penelope Hardy
Text Permissions: Neil Ryder Hoos

Library of Congress Catalog Card Number: 92–71193

ISBN 0–8109–2873–6

Published in 1992 by Harry N. Abrams, Incorporated, New York
A Times Mirror Company

Printed and bound in Italy by Editoriale Libraria, Trieste